ReWild

ReWild

the art of returning to nature

Nick Baker

Aurum
Press

Brimming with creative inspiration, how-to projects and useful information to enrich your everyday life, Quarto Knows is a favourite destination for those pursuing their interests and passions. Visit our site and dig deeper with our books into your area of interest: Quarto Creates, Quarto Cooks, Quarto Homes, Quarto Lives, Quarto Drives, Quarto Explores, Quarto Gifts, or Quarto Kids.

First published in 2017 by Aurum Press, an imprint of The Quarto Group.
The Old Brewery, 6 Blundell Street, London N7 9BH, United Kingdom.
www.QuartoKnows.com

A catalogue record for this book is available from the British Library.

ISBN 978 1 78131 655 9
eISBN 978 1 78131 735 8

10 9 8 7 6 5 4 3 2
2021 2020 2019 2018 2017

Typeset by SX Composing DTP, Rayleigh, Essex
Printed and bound by CPI Group (UK) Ltd, Croydon CR0 4YY

Disclaimer

This book reports information and opinions which may be of general interest to the reader. The reader should not undertake strenuous exercise unless medically fit to do so and is advised to consult a medical practitioner where appropriate. Participation in any of the activities referred to in this book is entirely at the reader's own risk. The publisher and the author make no representations or warranties of any kind, express or implied, with respect to the accuracy, completeness, suitability or currency of the contents of this book, and specifically disclaim, to the extent permitted by law, any implied warranties of merchantability or fitness for a particular purpose and any injury, illness, damage, liability or loss incurred, directly or indirectly, from the use or application of any of the information contained in this book.

MIX
Paper from
responsible sources
FSC® C020471

To my girls, Ceri, Elvie and Mother Earth

Contents

'We need to preserve the wilderness and its monarchs for ourselves, and for the dreams of children. We should fight for these things as if our life depended upon it because it does.'

Wayne Lynch, *Bears: Monarchs of the Northern Wilderness*

INTRODUCTION:

Stripped Bear

EVERYTHING SEEMED so familiar. It was as if I was looking at all I surveyed through a lens that blurred the specifics. If I squinted and peered out through scrunched eyes, I could be at home in England. It was only when I opened them wide that I could see that the details were slightly awry. I was on a well-trodden footpath, devoid of plants and polished from regular use. I was standing under an alder, one in a thicket of wizen, dwarfish trees that enclosed the track in a dark natural tunnel, their branches stretched and interlocking like thoughtful fingers a couple of metres overhead, occasionally loosely enough to let pools of limpid light through, spotlights on what lay beneath.

At home, some five thousand miles away, I would refer to this sort of habitat as carr: a stunted woodland where the trees, adapted to the sodden soils, send their roots twisting and coiling into the soggy mulch of water and fallen leaves at their feet.

A couple of metres above the stagnant quag, reflected in the oily water puddling on the surface, warblers of various species, as difficult to identify as they are at home, furtively flossed the leaves and cragged bark for minute morsels of invertebrate life.

But when I picked up and inspected a fallen leaf between forefinger and thumb, or managed more than a snatch of a glimpse of a warbler, I found that they were different in form and feather. Saw-like serrations edged the leaves and was that a touch of black on that green warbler's head?

This new view was a world that was somehow amplified and distorted, an increased vibrancy and scale with a twisting of perspective. It was both my childhood Sussex wood and alien at the same time. Then a berry caught my eye, glowing like an LED light in amongst the dimpsy autumnal greens and browns, illuminated in a pool of escaped light.

It looked a bit like a raspberry. It followed the general specifications of raspberries, although it was almost certainly a different species to the well-loved *Rubus idaeus* – the European raspberry that inhabited my berry patch at home. Something was telling me to pluck this strange berry from its receptacle.

When the salmon-pink berry willingly parted from its core, I was able to judge, by smell and touch, how ripe it was and, despite the voices of cultural conditioning in my head – the voices of my parents, aunts, uncles, teachers and many others of authority in my life – telling me not to put fruits unknown into my mouth, I did. A much bigger authority told me it was fine.

It was good, so good. As the scarlet drupelets burst between my teeth, they released a natural sugar rush. It dawned on me that the sweetness was on many levels. I had satisfied in a very small way something deep within me. I had seen, felt, tasted and smelt as a real animal, as a human animal; my latent aboriginal self had assessed the situation and had told modern, conditioned me that this was fine. No books, field guides or websites were consulted. If there was a risk that I had got it wrong, then that, also, was surely part of this most primeval thrill.

Then, just before I could get carried away with the moment and stuff yet another berry into my mouth, another moment, slightly more pressing this time, rudely barged its way to the fore.

My guide in this wild place, which was so far from my real home, pushed me slowly and urgently back off the path, his hand on my chest and his hushed tones demanding my complete attention. My survival instinct was piqued, the mood had changed, and something was about to take all my thoughts and ideas and, in one instant, reorder them. Taste buds to terror in a split second.

All senses stretched as far as they could, the cones of my retinas grasping at any photon of light struggling through the thick-leafed ceiling overhead. My intuition was desperate for information that might give me some kind of a clue as to what was shuffling along the long, dark, gloaming tunnel created by the trees ahead.

Hairs were actually prickling on my neck, all neurons were firing. Then, barely perceivable at first, I heard it: a low rumbling, deep and visceral; a distant thunderstorm rolling closer. I could feel this force of nature approach through the quaking ground, through the air thick with moisture.

When my eyes finally focussed on the hulk and they were able to decipher the dark fur from the dour foliage in the twilight, I felt a primeval jolt, unlike anything I'd ever felt before. A direct connection with everything that was and is now. A genetic memory maybe?

A bear bowled along, passing us by just a couple of metres. It barely turned to register us: it had a place it needed to get to; a better berry patch than mine maybe? Food to turn to fat. The only thing that matters here in Alaska in the winter, in the world of a bear, is survival. A couple of humans didn't fit into this particular bear's pre-winter season preparation and it had other things to preoccupy its mind.

I, however, didn't. My head was filled with bear; how could it not be? I could see, hear, smell, feel, almost taste bear – every primeval link, every neuron I possessed, woke up in an instant. Was I scared? Perhaps for just a modicum of a moment I was, as my central nervous system rallied around to find sense. Did I want to run? Not really. I was truly 'in the moment'.

It was something my alternative (some would call them 'hippy') friends were always talking about, but which I had never really understood until that particular moment. I felt so completely and utterly alive. I had, for that brief moment, completely and utterly connected with the natural world around me in a way that made complete sense. For the first time in my life, I got it. In an instant, that bear taught me what wild meant and in a few seconds focussed my entire person and made sense of every natural experience I have ever had.

My epiphany had a bear in it. Kind of appropriate, given its cultural and symbolic significance among peoples who still have an intimate relationship with nature. Many North American cultures consider bears to be spirit animals and they represent a strong grounding force, pulling us back down to the soil, keeping us in touch and in balance with the earth and with each other, as well as providing us with strength and courage to stand up against change and guiding us towards physical and emotional healing.

Some say 'bear' holds the teachings of introspection. When it shows up in your life, pay attention to how you think, act and interact. The bear is considered to be an animal that forms a bridge between night and day, strength and peace, the spiritual and the physical.

This 'my bear' had provided me with a sudden insight into a deep-rooted connection with an ancient wisdom. It had rocked me to my core and from that moment on my relationship with nature of any kind changed forever.

I had experienced the penumbra of this feeling before. As if the fingers of nature and my wild side had reached out and never quite touched, as if my conscience hadn't quite grasped the wild in my life. As a child I had walked in an East Sussex wood at night, unaware of the fact that the landscape was a man-broken, tame shadow of its former self.

In my eight-year-old head, the anthropogenic nature of all of my nature hadn't yet registered.

I still had that primeval fear of the dark; the bears, wolves and big cats of my imagination still roamed real. The alarmed explosion of pigeon or pheasant, the thud of rabbits' feet, or the grunt and crash of a fleeing badger would tear through my comfortable present and my racing heart and piqued senses would take me back to a way things used to be in some other life. What I had in these moments was something that every eight-year-old human ape had experienced for the last 7.5 million years or so. Ironically, to the animals I unwittingly bumped into, I really did represent a species of threat; they were still keyed in to that original wild wood, not the man-gardened Weald of farm, cover crop and cramped rivers.

The rest of my all-too-brief stay in Alaska's Katmai National Park and preserve was full of what I can only describe as a reawakening. Here I was exploring an ecosystem that was much closer to being complete than the one that remains in England, and while the species specifics were slightly different, the overall flavour was the same, the same rules of the game applied. That game of survival played by beaver and badger, elk and crow, chickadee and chaffinch was being played out with the same intensity on the other side of the world even if, back home, some of the pieces were missing.

That bear moment was one of many, and some encounters were even closer; at one point, I had a face-to-face with a huge thousand-pound male grizzly that had first walked, then run, towards me to check me out on an exposed flat of estuarine mud. I had had nothing but a bin liner in my pocket (a recognised anti-bear device) with which to defend myself. All of these experiences, and others involving the alarm calls of warblers, the unidentified rustling thing in the thicket, the estranged cry of something in the night, thrilled me to my core for the next week or so.

What occurred to me during this time was how much I noticed things, I mean really noticed things. It was as if the blinkers of my domesticated upbringing, together with the sensory curtailment that it created, had been lifted from my eyes and all my other senses had been similarly liberated. There is no better way of focussing on your environment – the lay of the land, the plants, animals, sounds, smells – than if your very life might just depend on it. As a naturalist, I have spent a considerable amount of time sharing my self-confessed biophile's view of the world with others. My career has spanned several decades of wildlife broadcasting, writing, guiding and outdoor informal teaching. During this time, I have often been told (and I mean this with as much humility as I can muster) how much I notice.

'How did you see that caterpillar?' 'How do you know that's a treecreeper singing?' 'Wow, you just spotted an orange-tip egg!' My clients seem to be in awe of these abilities. I see it in their eyes, I hear it in their whispered utterances. However, the skills that enable us to spot these details in the world around us are not difficult to acquire. In fact, we don't have to acquire them at all; we already possess them and have done since birth. We are all born naturalists.

The toolkit that enables us to be aware of and survive in a wild world is something we are all gifted with. It's already in our nature, but it's our nurture that distracts our innate ape. The twinkling baubles and glitter of our own cocoon of technologies fills our senses. Our neurons light up the switchboard with poor facsimiles of those wild connections to nature, our nature. In this displacement of the laws of the wild with those of our own making I feel we are leading ourselves up a garden path; one that leads not to some utopian perfection but that, ultimately, will lead to perdition via disaffection, distraction, disengagement and all manner of disturbance, for which there is plenty of evidence. Increasing rates of mental and physical health disorders, as well as other diseases linked to an increasingly sedentary lifestyle, are all part of this divorce from ourselves, but more about that later.

I guess I have my forward-thinking parents to thank for placing me in the countryside at the perfect age for the innate naturalist in me to develop. The countryside around my home had all the allure and excitement I needed to exercise and stimulate my body and mind. It became as much a mentor as my folks and close family members. It became a place of learning as good as any school classroom and, in time, it was my teacher, friend, gym, playground and, unwittingly, my therapy. I was already a biophile, and while not yet fully fledged, my natural fascination with the wild world was kicked up a gear.

Our first family home had been a new-build housing estate not far from Crawley, where there was no chance of seeing a deer or a fox. But here in my boyhood rural idyll these animals and more became things that I could realistically encounter outside the pages of the Hamlyn encyclopaedia of the animal kingdom. The early 1980s were an age before home computers and digital distraction. My parents, in an uncharacteristic lurch towards modernity, invested in a BBC Model B home computer under the auspices of it being 'educational'. Even in the days of blocky graphics and big pixels the addictive lure of computer games raised its ugly head, but my parents quickly instated rules and time limits.

However, I found that the embryonic, cold and slow-to-respond digital world had little to offer compared to a lubricious handful of frog spawn, the buzz of climbing forty feet up a tree to peer into a crow's nest or the dark but delightful symphony of witnessing a dead thing slowly being dismantled and returned to its elemental constituent parts. All of these experiences were real and engaged with the real me. As I practised, not in the sense of forced extracurricular piano or flute lessons, but naturally and in my spare time, I honed my sensory toolkit.

Bit by bit, lesson by lesson, running wild in those fields and woods, I was getting deeper and deeper into a world of thrills and skills that I have been using and further practising to this day. I remember watching the film *Legends of the Wild* (a fictitious story of an American mountain man wrongfully accused of murder and fleeing to the hills where he learned to survive and befriended various animals) and being completely seduced by the ideas and imagery in it.

The beguiling freedom and relationship Grizzly Adams had with nature was something that resonated with me. A bit of a loner at school, I felt I could never really be me in the regulated confines of the classroom, and while rural Sussex was a little short on large furry ursids, it did at least have badgers. As soon as the film finished, I remember getting dressed in a sheepskin gilet my nan had brought back from a coach tour of the Western states of America – the nearest I had to buckskins – and strapping a sheath knife to my belt. I was out of that door, crawling through thickets, wading across rivers, trying to get close as I could to, and to establish a relationship with, the wild things. This adventure was and still is my life. Back then it was what excited me, it was what I wanted and strived for, and it was what I held in highest regard. A relationship with animals and, by default, the rest of nature is what I valued and, by definition, it's what I excelled at over time. That's key – you get good at what you hold precious.

It's a lifestyle thing. While you might not be able to notice, recognise and see any rhyme or reason for the need to distinguish the sub-song of a robin, you might have other skills which, to me, might seem equally bizarre. Some of us can hear, recognise and even date a train just by the sound it makes: the collective of rattles, metal-against-metal rhythm of the bearings and pistons and combustive breath all come together to form a symphony as identifiable and distinctive (to those who listen) as any philharmonic orchestra or, indeed, any band, musician or instrument. What you are tuned into really depends very much on what it is that you give personal value to, and while these particular examples all refer to sound, the same principles can be applied to pretty much any of our senses. You might be able to taste the phenols in a good single malt, the hint of chocolatey tones in an Australian Shiraz or the homeopathic dose of chilli in a salad oil.

I could end this book right here, really, as none of this is rocket science. I look around me and I see a world populated by smart apes that have forgotten where they came from, running around in circles, unhappy and dysfunctional, and that is because that is exactly what we've allowed ourselves to become.

The fact of the matter is that you need to want to do it, you need to value that connection with nature in some way. So if you want to run wild, understand the world better, get to know your own personal ecology, get fit in body and mind, become an ace birdwatcher or award-winning wild-life photographer, or simply to feel a real sense of satisfaction, not just an urban, disconnected buzz of getting a two-for-one deal on prawn crackers, then read on.

1

Defining the Wild

T HE POPULARLY perceived meaning behind the word 'rewild' is a kind of muddled-up, querulous, polarised and politicised soup of ideas: beavers, lynx, bears, wolves and George Monbiot all served up with a confusion of tabloid headlines where beavers eat fish, bloodthirsty lynx will consume all the sheep and elephants will restore the balance of nature (well, actually, there is some truth in the latter).

But what, exactly, is rewilding? There's a lot of talk about this word right now and to those who don't speak the language it can seem rather confusing. There seem to be many different kinds of rewilding too: there is cultural rewilding, landscape rewilding, personal rewilding, even Pleistocene rewilding! It's clearly a word that has many different definitions. It is the more exciting, controversial and sensational forms that dominate popular culture and these seem to touch some deep-set cultural nerves as well as some long-suppressed inveterate guilt and embarrassing truths. It's often an emotive, contentious word laced with prejudice and opinion.

The verb 'to rewild' most often refers to an argument that posits that large predators and other keystone species are integral to maintaining the integrity of the ecosystems in which they would naturally exist. Such species have an influence on many other species below them in the trophic food chain – this is something called an ecological cascade effect, a top-down stimulation of what ecologists call trophic diversity. In short, this is the number of opportunities for different species to eat each other and therefore support their own. A good example would be the reintroduction of wolves to Yellowstone National Park. After being absent for over seventy years, wolves were introduced in the mid-1990s to a land that clearly desperately missed them.

When the wolves come back, they started an ecological cascade, influencing not only the populations of those creatures that they hunted, but also their movements and distribution which, in turn, had all manner of knock-on influences on many other species as an indirect consequence. Deer numbers were reduced and the deer stopped becoming lazy creatures of habit – a predictable deer is easy prey for a predator. A wild deer in an intact environment is always looking over its shoulder, is always on the move, a nomad driven by the need to feed but equally to not become food.

Prior to this, the deer in Yellowstone lingered in favourite spots and annihilated the vegetation. Trees, bushes, plants and herbs were nibbled down and plucked to nubs by the teeth of herbivores.

However, when the wolves arrived, they stirred up the trophic layers that had settled into a dysfunctional version of their pre-human existence. The plant life, for example, relieved of the constant barrage of cloven-hoofed beasts, started to recover – the trees grew, the grass grew; there was plenty for the insects to feed on; then, suddenly, the warblers had caterpillars to catch and trees to nest in; the shading of the water in the rivers meant that invertebrate life flourished here, too, and therefore fish numbers and diversity went up. Put the wolf back and the birdsong gets louder and the angling opportunities improve – it's actually way more complicated than this, and raw stirring is still happening, but you get the idea.

'Rewilding' is all about protecting what is left of natural and semi-natural systems and improving lost functions. Reintroducing those species that are missing is key to this. As is giving them the space they need, or, rather, giving back the space they need. It is about extending and linking together core wild areas in a landscape scale process which allows larger animals the extensive space and the connectivity they require to flourish and be ecologically resolute, especially in the face of change. This seems to be something we've got wrong.

In short, for the most part, the definition of the modern rewilding movement is an admission that the existing model of nature conservation hasn't worked particularly well as a complete solution.

Adding nature or subtracting culture

Picture a scene: a blood-red rhamphotheca pops up against the verdant-green sward; the two-tone sanguine eye watches all. Other than the barely perceptible flicker of an eyelid, and the wind-blown ruffle of feather, she's motionless.

Turn the sound off, and there seems nothing very much wrong with this scene, it's a colour-wheel lesson in complementary pigments. The oystercatcher is on her nest, sitting on four ovoids of granite grey, blotched and blotted with brown. These eggs are her everything, her polestar; they may be her only contribution to the next generation. Not that you can see them, she's stuck down tight, hunkered on a simple nest cup, protecting and nurturing the seeds of life within, keeping the herring gull's gaze and the chill Scottish breeze from their matt surface.

Pull back from this microcosm of spring tranquillity, past the bobbing thrift above the rock-rose, past the rabbits keeping the sward tight and bouncy, past the white-painted line, the crisp packet and crushed can, the tarmac, the splintered bone, brain and fur of a baby rabbit, then turn the sound on: an unremitting din, the dissonance of motorbike, bus, lorry and car, 24–7 noise and grunge.

This is modern wild. A wilderness in the round, a fragment of the Highlands on a traffic island in Inverness.

It works for now, for the thousands of us who drive around it and for those with their feet placed on its soil. It's a fragment of something that feels right to the oystercatcher and rabbit, and countless other wild species which undoubtedly share this atoll on the A9, but for now remain unseen. They're safe from predators with all the facilities a nesting oystercatcher or rabbit could possibly require, until they need to leave, as someday they must.

Initially, this scene struck me as nature fighting back, life finding a way. But not far away, in the glens and straths, oystercatchers just like this were going through the same life cycle. This was not a case of birds having moved in; rather, it was that we had moved in on top of them, our lives super-imposed on theirs without any consideration for their needs. Obliviously, we had carried on, laying down the tarmac, expanding and spreading, covering over the landscape.

It seemed incongruous and uncomfortable: a beast of the wild wind-swept places, being buffeted by the draught of trucks, dust and diesel fumes. But while this situation slapped me in the face because of the extreme juxtaposition of its elements, it is what it reminded me of that was most poignant: nature next to, but separated from, culture.

A similar scenario is unfolding all over our countryside and, for that matter, all over the world.

While the oystercatcher and the rabbits were happy in their moment, safe in their albeit noisy sanctuary, it bothered me to think of what might happen when the doddering chicks hatched and started tottering around, or when the rabbits, having bred like they're supposed to, needed to disperse. What was happening on this roundabout was a microcosm for how most nature reserves work (or don't work). A compartmentalised idyll, a fenced, walled-off fragment of how things once were all around it. The word 'conserve' means to keep something the same, to preserve the stuff you want, like jam in a jar. But while a sugary mush of fruit can sit on a shelf, preserved indefinitely, in nature, things don't work in the same way.

The main conceptualization of conservation since its birth in the eighteenth and nineteenth centuries has been all about compartmentalising the landscape, a desperate land grab to hold on to rapidly disappearing habitats and the species they contain.

For 99 per cent of our existence as a species, humans have been hunter-gatherers, a nomadic ape with a big head. A big head on a body that, ultimately, wanted to sit down and do a little less running around in pursuit of dinner.

Then about ten thousand years ago, someone started this thing we call agriculture – a process by which we could stop the exhausting necessity of chasing things around.

Since that time, as a species we have been on a tip. We have made finding food more predictable, easier. We've fenced in the creatures we want to eat, and stopped them running off and thereby eliminating the need for energetic pursuit and, similarly, we've worked out ways of growing and culturing the plants we like to put on a plate next to them.

This ambition to make life easier for ourselves has become a bit of an obsession, one in which we control and manipulate, mould and shape the natural world to feed our needs. To distance ourselves from the wild and our natural ecological state has become the human model of progress. The process of agriculture, first experimented with in the

Fertile Crescent of ancient Mesopotamia, has allowed us to settle down and ultimately create our cities and civilisations. This, in turn, has given us plenty of time to think and design things, new stuff to make our lives even easier and to enable us to overcome other problems we have inevitably caused for ourselves. The channelling of water, trade networks, roads and boats all quickly followed. This process of control and exploitation for our own needs ran uninterrupted right into the early part of the nineteenth century. We've spread out like a plague. Where before we slipped between the trees and moved through the grass, now we seek dominion; we've moved and removed them, trod on and crushed whole eco-systems as we've reshaped the world to our own needs and requirements to the detriment of nearly all other species and the habitats that support them.

However, there were some who wisely questioned this consumption of the land and a counterculture was born. In 1821, the English eccentric Charles Waterton built a three-mile-long, nine-foot-tall wall around his estate Walton Hall in Yorkshire at a cost of £9,000 in order to protect the wild birds and other wildlife that were dear to him from the proclivities of poachers. In doing so he unwittingly created the first nature reserve (this forward thinker is also cited as the originator of the concept of a bird box). Shortly afterwards, across the Atlantic in 1872, President Grant signed an order for the first national park and Yellowstone was founded in order to protect its unique geological, geothermal and landscape features from being exploited and spoilt. The conservation movement had started moving.

Traditionally, the nature reserve, national park and wildlife preserve have all been about sanctuary for nature, fencing it off and protecting it, excluding the species which has been having deleterious effects on it – namely us.

However, what started as a desperate measure to *save* places in the face of almost certain anthropogenic devastation, while working in the short term, has fallen short in the long term; quite literally, in some instances. While in some cases nature reserves have been useful repositories for vulnerable species and habitats, sanctuaries born out of desperation, the reality is that they are rarely big enough to work in the long term.

Yellowstone, with its 3,500 square miles of wilderness, might seem big enough to function naturally, but this couldn't be further from the truth. While it seems big to a human who might wander a few miles a day, or to one who grew up in a world with critical bits missing, for an individual of a species such as a wolf, bear or wolverine it's an island.

Sure, it's one that works for a while, but just like a machine with countless working parts, without mechanical maintenance the thing eventually grinds to a halt, all those intricate mechanisms that we are only just beginning to understand need looking after. Wolves or beavers or any other keystone species are the ecological engineers; they have their basic requirements and if they are not met and are lost from the system, sooner or later that system fails to function properly, leaving nothing but a disparate collection of species, all out of whack with each other. It's what is often termed, rather over-simply, as the balance of nature, unbalanced.

If wolves or beavers can't meet others from neighbouring populations, something which is as necessary to the long-term sustainability of their species' population as enough habitat and food is to their short-term sustenance, then it's game over. When a wolf, bear or bison leaves Yellowstone, it inevitably meets a human and is shot. Think back to the fate of the fledgling oystercatcher chicks tottering on the edge of the roundabout: it's different in scale, and there is a human motivation in the latter, but the reason for moving and the outcome are exactly the same.

This bigger picture is something that until relatively recently has escaped us. A pot-plant or Noah's Ark philosophy has reigned in our relationship with animals and plants; it's a kind of two-by-two mentality, one species at a time. The traditional zoological or botanical park is how we've frequently seen conservation – some of these establishments call themselves conservation organisations. But even these captive populations can't exist in isolation – that's what breed registries are all about, maintaining, as best we can, some kind of genetic integrity, as far as our somewhat arrogant understanding of such things goes. Not all that long ago, we thought the 3,500 square miles of Yellowstone was plenty, but in the meantime technologies have allowed us a greater understanding of the dynamic nature and the extensive spatial requirements of many large-animal populations.

Pluie was a famous example of wolf kind. She was five years old when she was captured and tagged. In just nine months, this radio-collared animal covered an area over ten times that of Yellowstone, some 38,000 square miles, and crossed three state lines before she was killed by a hunter. M65, a similarly radio-collared young male wolverine, travelled over 500 miles and through at least four states before he, too, met his demise at the hands of a man who thought he posed a threat to his livestock. These are large and obvious creatures, and as long as they don't end up in the crosshairs of a rifle or meet the grille of a truck, they will find a way. They'll seek out spaces that work for them, they have a natural predisposal to disperse, to roam until they find a vacant territory, new potential mates and, of course, the food resources to sustain them. Thanks to radio telemetry and GPS systems, we're beginning to realise the failings of our previous understanding. Many of these large animals – bears, wolves, cougars, wolverine, elk and moose – are capable of clocking up the miles in search of dinner and a date.

Like a bad joke, or a repetitive nightmare, this exact same situation is repeated almost anywhere you decide to look on the planet's surface. Large-space, hungry animals are confined to increasingly fragmented habitats. They are fenced in, sometimes literally, sometimes by boundaries invisible to us such as green deserts of arable land, open pastures, roads and developments. Tigers, jaguars, elephants, lions, polar bears, maned wolves, giant armadillos, take your pick. Look hard enough and you'll find conflict. There is no space for these ecological giants.

However, at least many of these animals have fans, defenders of their rights. It's easy to stick an ape on a poster or in a magazine advert and generate sympathy for their plight. The power of wet, watery eyes is not to be under-estimated. That's why we are culturally aware of them – but what about other, smaller species that are just as important in their own way? They may not be keystone species but they score the same when it comes to counting the diversity of life, the very thing that makes our world interesting.

Have you heard of Desmoulin's whorl snail? Unless you remember the headlines surrounding the 'Battle of Newbury' in 1996 and the construction of the Newbury bypass, the chances are that you haven't.

This little three-millimetre-long brown twist of mollusc life halted development of a major road and was literally picked up out of the way and moved elsewhere. Conditions for its survival were not quite right, however, and it died out: sad but sadly familiar.

In the UK, despite the fact that we have 15 national parks, 224 national nature reserves and thousands of others owned or maintained by non-governmental conservation organisations, and protective legislation, there is an overall depressing failing in the initial intent. We are still haemorrhaging biodiversity at a dismal rate and failing to reach any of the targets we set ourselves, so our current model is clearly not working.

Nature reserves, in whatever form, are not doing their fundamental job and, just as in North America and other parts of the world, the growth in human population and the continuous roll-out of our civilisation and its trappings, such as habitat fragmentation, building and developing cultural infrastructure with no mind for nature, are responsible.

Life, whether in the form of wolf, wolverine, beetle or butterfly, aconite or aspen, struggles to exist when its populations are distanced from each other. It may limp on for a while, but ultimately it loses its genetic diversity and also its robustness to environmental change. And, boy, is it changing – climate change has been identified by some as being the single most important challenge humans, and by default the rest of life on Earth, has ever had to face.

Many animals and plant species exist as part of much bigger meta-populations, which, while appearing to exist in isolation from each other, still depend on flux – some degree of emigration and immigration between populations in order to keep them robust and healthy. Our cities, houses, roads, highways and other developments, our bricks, mortar, tarmac and monoculture, come between these fragmenting populations. They lose and so do we. In short, they don't work, they're too small to be sustainable. In situations where the key species have gone missing, for whatever reason, the biospheric loss and ecological fallout constitute a slow demise of the entire system, species by species, as the ecosystem breaks down, falls apart and senesces.

But bear with me. While this may seem a little heavy and make you want to give up on me, this book, the human race, even the world, please don't; there are plenty of reasons to be cheerful and that's what I'm leading up to. One of these is something called Y2Y, a catchy acronym for the Yellowstone to Yukon Conservation Initiative.

This brilliant partnership of some three hundred individuals and organisations from both sides of the US/

Canadian border was started way back in 1993, in recognition of the very things we've been talking about.

Y2Y stretches some two thousand miles from the Yukon territory on the edge of the Beaufort Sea and well within the Arctic Circle down through the Rocky Mountains to its southern outpost, Yellowstone National Park. The zone takes in some of the last remaining regions of American wilderness. It was set up in recognition of the fact that as well as being as close to bio-intact as anywhere, with a full suite of large herbivores and carnivores and all the ecological minions that prop them up on their shoulders, it was also well endowed with protected areas – some forty-four national parks in total. These parks, while acting as useful cores, were simply not big enough for the sorts of natural peregrinations of the wildlife.

To witness first-hand the dynamism of nature is something difficult to put into words. When I saw tens of thousands of caribou of the porcupine herd from a distance, they looked as numerous as the mosquitoes attempting to drill into my clothes. Yet each of them, weighing some 200 kg, was supported by this seemingly barren tundra. As I got closer, I realised they were moving for the same reason as I was flapping my arms around and clawing at my face. Mosquitoes were hounding them too – and feeding on the mosquitoes were many birds of different kinds from perky little waders, nipping between their feet, to wagtails and dashing jaegers.

I had been hanging out with Karsten Heuer (a wildlife biologist) and Leanne Allison (environmentalist), who, in order to draw attention to the importance of Y2Y and as a bit-part protest against potential development of the Arctic National Wildlife Refuge – the core calving ground for the herd – had lived with and followed 120,000 of these animals on foot for a year. They had travelled over some of the most testing terrain in the world for some 900 miles, sleeping when they slept, moving when they moved. They had seen and lived it all; they were twelve-month honorary caribou.

They told me that their main realisation was that it wasn't just the mosquitoes and the birds that relied on this moving maelstrom of deer flesh and blood. It was like an ecological procession. The caribou 'sucked an entire eco-system along with them'. If it wasn't bears it was wolves, continuously harrying the herd, plucking the weak from the strong, the chaff from the wheat. This 900-mile journey made by around 197,000 animals is the longest migration of any land mammal and is the intact northern equivalent of what I was to see at the end of my journey south.

In Yellowstone, all those that remain of the great bison herds of the Midwest stand around in clusters. I had wriggled my torso out of the sunroof of my rental car to watch these gargantuan cows, stoically resisting the biting winds, snow-flakes catching in their hair and eyelashes, both adamantine and delicate at the same time. Having paid my fee, nature felt like a commodity. It was necessary, of course; the ecology and the economy, and keeping both human and environment as healthy as possible, had meant that the two had become inextricably intertwined. Here nature had a value, while up in the north, other than to a few populations of indigenous people, the caribou were more of an irritation to the petro-chemical companies wanting to exploit the fossil fuels beneath the calving caribous' cloven feet. Two extremes of the same relationship. It was while making this five-thousand-mile road trip (the irony of the fact that I was driving didn't escape me) from north to south that, effectively, I travelled in time, from the near-pristine wilderness, with a single Dempster highway, to one in which the oppressive influence of strip malls and convenience stores and fast-food franchises represented the way we live today. On the way I met indigenous people, modern wildlife managers, hunters, artists and conservationists, all of whom sang with hope about the ambitious Y2Y conservation initiative.

There is hope on both sides of the fence, quite literally. The Trans-Canada highway is busy. It's the main route between Calgary and Alberta and over fifteen thousand vehicles travel the four lanes every day. Near Banff a twenty-eight-mile stretch of it is fenced on both sides with a 2.4-metre-high fence, to reduce the high numbers of road-traffic accidents involving large wildlife. While this is of benefit to humans and to wildlife in the short term, the problem again is one of fragmentation. Although the road was always a barrier of sorts to wildlife ever since the highway was built in the 1950s, some animals did manage to cross. The fence, however, made it a barrier impermeable to even the most determined moose.

But what encapsulates the Y2Y model is that a forward-thinking blend of integrated science and education has been employed to build a network of underpasses and overpasses with the aim of restoring the population connectivity on either side of this busy and critical transport corridor.

As I stood amidst the head-high conifers, dense scrub and the tangle of knee-high grasses and herbs, it was hard to believe that I was here. According to my GPS, I was slap bang in the middle of the highway; there should be four lanes to the north and another four to the south, but here there was the occasional scratchy birdsong and the singing of the wind in the pines. It was as if the landscape had slid down off the hills like a loose stair runner and engulfed the Trans-Canada highway in its lush, verdant shagpile – a naturalist's eco-fantasy in reality.

My map reading was fine and my GPS was fully functional – it didn't lie. I was exactly where it said I was, only above it.

If I concentrated when the wind died down briefly and the bird stopped its somewhat tuneless vocalisation, I could make out the gentle purr of the traffic. This is exactly the point. I was right in the middle of the fifty-metre-wide wildlife overpass – a short corridor connecting wild habitat on either side of the deadly black top. This was one of two like it on this stretch of highway; the fact that I couldn't see any edges, hedges, walls or fences was so that even the most timid species would make the journey.

The Banff Bow valley overpasses are exemplary wildlife corridors, and spectacularly visual at that, but they are just one tiny example of the ethos behind the Y2Y vision. It's not about segregating wildlife and people, it's about integrating and connecting so that both can live and thrive side by side.

This, for me, was touching the future and, while there is still need for rangers, wranglers and intervention when wildlife gets cornered, overall there is an awareness and sympathy for the wild, and the flavour it brings to everyday life is appreciated. Even a victim of a near-fatal bear attack rather nonchalantly told me, 'I was on his turf, right; he was just doing what bears do.'

While this isn't so much rewilding as reconnecting, it does resonate with what this book is about: it is our relationship with nature and how we must re-evaluate our place next to it that is so important.

2

Starting Small

A LOW-FLYING COMET, a rocket of red, a will-o'-the-wisp, the squirrel raced helter-skelter through the Scots pine wood. It was there one moment and gone the next, up and down, in and out of the greying pillars of pine. There was never more than a very short split-second view, but it didn't matter to me that I didn't get to see its perky form, those iconic tufted ears and sparkling, untrusting eyes. Just a glimpse of this, our native British squirrel, was enough to satisfy me. But it shouldn't be. For a start, I'd travelled nearly seven hundred miles to see one, and this was, after all, *our* squirrel. Mention 'squirrel' today in the UK and the animal that springs to mind is exactly the same one that would spring to mind if you asked someone from North Carolina. The one we call squirrel is now grey, not red.

This is an example of what is known as 'shifting baseline syndrome', and it is something that comes up quite a bit when you're trying to work out the true functionality of a British ecosystem. The problem is that we're quick to forget. What one generation grows up with becomes the norm, becomes the target we strive for. If my parents had got their act together and had had me just a few years earlier, it's possible that I would have been seeing colour in my local woods too. That's how close they were, how quick the change was and how short our collective cultural memory is.

This insidious, slow, creeping change is just one example of how our own perception of what is wild alters with time. What belongs and how the land functions gets tangled up and creates a rocky ground in which to sow the seeds for any rewilding.

One question that is helpful to ask ourselves is where exactly are we aiming at on the time-line – from the first Cro-Magnon footfall on British soil some eight hundred thousand years ago to 1988, when we lost the short-haired bumble bee?

There are countless other examples of this phenomenon of shifting baseline syndrome. The further we get from the land, the less we talk about it, the more ignorant we become as a species and we lose our way. Separated from the mother that ultimately nurtures us all, we struggle to understand because we've lost sight of what is real and what isn't, we disconnect and nature-deficit disorder follows. We are walking blind in a land of shadows.

I have heard quoted the 'fact' that there are now *too many* sparrowhawks, buzzards, white-tailed eagles, foxes or bad-gers numerous times. In fact, there's a club where simply being in possession of a hooked bill, sharp teeth, claws and talons gets you guaranteed membership.

Furthermore, our own species has a habit of ecological xenophobia which applies to pretty much anything that seems to have a will of its own. Many people seem to want to 'control' and eliminate nature from their garden, to hold dominion over every living thing, big and even very small. I knew of someone who wanted to eradicate the mining bees from her herbaceous border, for no other reason than that they were making it look a mess. This short-lived, innocuous animal would give me great delight in my garden and would actually be of immense benefit by delivering pollination services. For those of you not familiar with a mining bee, it's a small, foxy red-brown ball of winged fuzz no bigger than your little fingernail. It's a solitary bee that likes company, an apparent oxymoron meaning that it nests on its own, but it likes to have neighbours nearby. While the nest itself is a short tunnel as wide as the bee, the excavated matter kicked out by the industrious insect forms a small, loose cone of fine soil, no more than a centimetre or so high. It's a spring species and is usually on the wing for a short period of time between March and May. A perceived need to get rid of such an insect is, to me, further evidence of how as a nation we've lost the plot and become ecologically intolerant.

Just mention the word 'bee' and everyone is either reaching for a fly-swat or an EpiPen. The fact is that most of our 230 bee species are solitary and, because they don't store honey that needs defending, they aren't aggressive and don't pack any form of venomous punch. You would have to try very hard to be stung by one, and by that I mean you would have to pick it up and squeeze it and even then most cannot do a thing about it as their hardware can't even penetrate our skin.

It occurs to me that if we cannot accept a small, passive, positively helpful pollinating insect in our flower beds, we are probably not ready to embrace the reintroduction of mega-charismatic, long-missing, keystone species such as beaver, wolves and lynx.

Following on from this thought, the question is, why not? Rewilding has become associated with such mega-toothy and charismatic species, but strip it back to its most basic defining action and the principle of rewilding can be as simple as just letting the corner of your lawn grow up a bit longer than usual. Developing a tolerance for the odd daisy or dandelion pushing up through the tarmac or letting some moss grow in the cracks in the pavement is, at its most fundamental, the start of the rewilding process. Or is it?

It's the start of allowing the wild to manifest itself in your own space, giving you the opportunity to immerse yourself in the poetry of nature. It's the kind of process that is at the heart of landscape and habitat restoration; you create the habitat and those animals able to find it will, in their own time, move in. It is a kind of passive rewilding, taking your hand off the lawnmower, spade and trowel and letting nature just seep back of its own accord – in all its tendrilled, stilt-legged, gauzy-winged, velutinous, feathery, creeping, budding, flowering fecundity. A rewilding purist would say that to rewild is simply to leave alone.

If you've dug a pond, put up a bird feeder or nest box, you're taking part in assisted rewilding, making a habitat more desirable for other species. Some would argue this has little to do with the wild and is a form of gardening. However, right now as I write this, a blue tit has joined another on the feeders outside my window. It just fanned its wings (the avian equivalent of growling) at another seemingly identical blue tit on the peanut feeder and now there is only one. This bird is as wild as any tiger and he (or she, it's difficult to tell unless you're a blue tit) was moving through the garden, going about wild behaviour as he or she might be doing in the ancient woodland just across the valley.

The next logical step would be the equivalent of reintroducing something that has gone missing, obliterated from the habitat that it was once part of, especially important if that species is a keystone species or one that creates the flavour or captures the ancient character of a particular habitat.

If you plant any native tree species, or some wild flowers once found in your locality, or introduce some frog spawn because there were frogs hopping around in your borders once but not now, then what you're doing is pretty much the same as those proposing to let a handful of beavers into a strath in Scotland, or lynx or wolves. You're just doing it on your own patch in your own way.

A good little cerebral exercise is to picture what the world would be like in your street, if everyone did a relatively simple thing and planted a native tree. Imagine if everyone in Croydon planted an oak tree in their garden (ignore, for now, the practicalities of roots and sewer systems, gutters and drains clogged with leaves, etc.) and just left the oak and the garden to do what they are genetically programmed to do: you'd have an urban oak woodland and all the joys that go with it in a couple of hundred years – maybe some roe deer would move in, you might have nightingales singing on every street and, if you're lucky, a lynx might find its way into your garden. True, not everyone desires all of these things, but imagine a world without any of those experiences to be had anywhere, because that is probably where we're heading.

Obviously, developing this scenario to its conclusion would entail everyone giving up on the garden and, of course, someone would have had to reintroduce that lynx (and all the other missing flora and fauna that had evolved to function together). We would have to resist the urge to tinker with, and remove, things that we didn't find useful or that we might consider harmful or simply no good. After all, this is human nature, it's what we do and have done since we ourselves left the trees. This brings up another factor: us. Surely we are part of this system too. Human nature? The natural human would have been part of this ecology too. At what point in our history did our relationship with nature go wrong, to the detriment of everything else?

You can see that rewilding is a scale. You can plant a native species of tree, local to the area, and you've made the place a little bit wilder than it was beforehand. If you add some native wild flowers and a little scrub, then make it possible for every species that would have existed in your patch before our anthropogenic influences spread throughout the land to wander by, resist the urge to tamper and watch what happens once you leave nature to its own devices, then you've got the other end of the scale – rewilding at its purest.

Wherever you come on this scale, rewilding is an acceptance of nature and of nature knowing best, a green wisdom, the true value of life. While it might be a bit of a stretch to imagine a Britain with wolves running wild somewhere in the landscape, it's not so difficult to think of one with a few more salmon in our rivers, or for that matter, frogs in our ponds.

It's true we've lived a long time in the UK without any predator bigger than a fox or badger, but if we can get into the right frame of mind as a nation, and I appreciate this is a big ask, then anything on this rewilding scale is theoretically possible.

Deadhead

What is it that makes me get the mower out and massacre the daisies and dandelions in the summer? Why do I deadhead the flowers? Yes, like a lot of us, I do from time to time. Despite this, I will spend many moments in the garden peering at the tiny denizens that occupy these plants and the micro-habitat they contain and, if the truth be known, I probably spend more time and derive more pleasure from these animals than I do from the deliberately planted flowers.

The reason I start to push that cylinder blade backwards and forwards as soon as the grass starts to grow is simply social conditioning, it's what I have grown up perceiving is the norm, which is the reason we struggle as a species and become the hypocritical 'Nimby', happy to campaign for beavers to stay in our rivers but unwilling to allow the grass to grow in our own backyards.

Rewilding nature is about joining up what is left of the shattered habitats that once were and about reintroducing species once seen as competition or threat in order to re-establish a natural order. The same can be said for us humans: we've become separated, distanced from our natural selves, and there's a desperate need for us to reconnect. We've become unbalanced, disenchanted, lost in a shadowland of our own creating both inside, perceptively, and outside, ecologically. I believe rewilding is a big word that can change our relationship with the world.

While 'rewild' is a powerful and multifaceted concept, it is much more relevant and closer to us all than we might at first think. Rewilding your attitude to nature, culturally appropriating its qualities, could just as easily be described as 'rethrill', 'rezest', 'resense', 'reincentivise' and 'rejoy'.

What links the plight of the oystercatcher with a wolverine, a bee with a beaver, is our relationship with nature; it's how we think of it, how we perceive it and feel it in our hearts that matters.

It's clear that the process of rewilding presents a number of theoretical and practical challenges and while the long-term aim of reintroducing some of the major missing players in our current ghost lands is vital, in reality this situation is a long way off. If we are to succeed in this and in the restoration of large swathes of landscape, we need to train ourselves to become much more tolerant of the wild.

This is much more than just getting beyond those tabloid headlines, where the mere mention of wolf, bear or beaver sends everyone into a tizz. There is no doubt that we need these 'big headline acts'; of course we do. These are the keystone species; we can never restore complete functionality to any ecosystem unless they are in place.

They are the mascots of the word 'rewilding' and stand to remind us about the bigger goal, the ambitious aim. We cannot have the ecological cherry on the cake if we've not got the cake yet. We've still got a few crumbs of the original functional landscape in many places both in this country and around the world, and while we still have crumbs and all the ingredients, there is hope. We can still bind them all together and rebuild and rewild our world, but first we need to reconnect, to re-establish and recalibrate our valuing system. We need to effect a change in the way we look at and interact with nature. Not as something that always needs to be tidied up, compartmentalised or controlled: we need to start integrating it into our everyday lives. The great thing about this kind of rewilding is that it is something we can all do. It's more accessible than you might think.

There is very little point in coming to terms with a wolf or a lynx running through your local woods if you struggle with the idea of a fallen tree looking untidy, a bit of mud, letting the corner of your garden run wild, letting your lawn be unmown or, indeed, tolerating mining bees in your herbaceous border. But to understand why this is important, you first need to look at how you view nature. What is your take on the other species that you share your space with? Do you even notice them? This is the sort of rewilding this book is all about. It is about rewilding you, rewilding your life, rewilding your attitude, rewilding your mind, while rehabilitating your heart and soul – and, in doing so, starting a process of deep satisfaction, love for ourselves, other animals and the place we call home. It's a natural therapy and antidote to the modern, technological bubble we race around and around in.

Where do you start? That's easy: you start with what you can do something about, and that is you. Somewhere inside is a sensitive creature, one honed by seven million years of hominid evolution. You're carrying with you right now all you need in order to engage with and experience the natural world, as well as other humans alive today. You are bristling with sense organs through which you can gain some amazing experiences which will give you a deeper comprehension of both yourself and your natural environment. This book is not an instruction manual that you need to follow religiously; rather, look at it as the helpful salesperson who has one of these at home, likes it quite a lot and wants to share with you its best features and how he has come to love, appreciate and value what it has done for him.

There are plenty of resources out there telling you how to do this and the best way to see it. I should know, I've written a few of them. They're great at one level, but what they can't do is take the place of the real joy – of the exquisite journey of personal discovery. A huge part of being a naturalist and connecting is what you learn and discover yourself, the sorts of things you don't get out of books. I spend a lot of my life answering questions and showing people things that amaze me; I can't help myself, it's only natural.

My big worry is that every time a question is answered, whether on TV, an Internet forum, in the social media or a school assembly, I deny the questioner that indescribable deliciousness of finding out the fact for themselves – the process of the hunt and the hard-earned fact, the 'kill' that stays with us; the exhilaration, the thrill of the chase and, of course, the undeniable truth of what is discovered. We live in a world of instant gratification, of lazy enquiry and a Google mindset; how refreshing is it to be left to get on with it, to find out for yourself? I say get out there and open your senses, witness nature first-hand, relish the details and complexities and revel in the interconnectedness of all things; marvel at them and enjoy joining up the dots for yourself. In the process you'll shake up an inner child and wake up the hidden monkey.

3

The Monkey's Eyeball

THANKS TO Aristotle and the other great thinkers of his time, if you ask someone in the street how many senses they've got, they will tend to answer 'five': sight, hearing, taste, smell and touch – this idea is very deeply ingrained in all of us, it's what we were taught at school. The five senses, of course, correspond to the five gross anatomical features with which we are endowed, and through which the bulk of our sensory input is received.

However, if you look – or, for that matter, use any other sense you feel appropriate – a little deeper, it doesn't appear to be quite as simple as this.

The five traditional senses that we think of today – sight (vision), hearing (audition), taste (gustation), smell (olfaction) and touch (somatosensation) – were derived from a much older idea.

In the medieval period, for example, we had our wits about us – ten, to be precise. It was believed that we have both internal and external wits. The external ones correspond to the five we're all familiar with and, in addition to these outward wits, we were also thought to have five inward wits: common wit, imagination, fantasy, estimation and memory. Common wit is what nowadays we might refer to as common sense, while estimation is what we might refer to as instinct.

As the phrase 'common sense' intimates, the words 'wit' and 'sense' in old English were very much interchangeable. Whatever your thoughts might be on this, it seems that having merely five ways of understanding the world might seem to be an over-simplification of our sensibilities.

Even when you take into account our contemporary understanding of our faculties, it seems that this idea of just five senses doesn't seem to stack up. We know for a fact that we have other senses, and not merely the inexplicable 'sixth sense', which has always suggested a sort of supernatural, extrasensory perception of a mystical quality. We can do better than that. Try a seventh, eighth, ninth and tenth sense – we have plenty of other sensory modalities at our disposal, some of which we're only just beginning to understand properly. We might need to get past the big five; it seems that by believing this is our lot, we might actually be holding ourselves back a bit when it comes to the business of sensory immersion in our living, natural environment.

Much of what you believe and understand at this point starts to come down to semantics – what actually constitutes a sense? Something that can't be disputed is that few of us really use the ones we know we've got to their full potential and so, since they are really all we've got to 'make sense' of the world, it's easy to work our way through the kitbag.

Let's look at the big five we know well. What they are, how they work and, more importantly, in the context of this book, how we can all use them to greater effect.

Smack, 'Ow!' came the sound of one child walking straight into a tree, followed by that of another falling over having tripped on a tree stump, then, after a short pregnant respite of no sound at all, a cacophony of wailing, moaning and the sorts of noises that usually accompany displeasure in the young. All rent the peaceful night.

My experiment at a night-time sensory walk not only sounded like a failure – when I turned on a torch, it looked like one as well. There were kids rolling around on the ground clutching their legs and arms in pain; some were lodged in bramble thickets unable to go forward or return to the path they had taken; others stood there frozen on the spot with varying degrees and symptoms of nyctophobia. The aim of the game, for that was what it was supposed to be, was to prove that there was nothing to be afraid of in the dark. However, this was also the exact moment that I realised my night vision was a lot better than I had been aware of. The question was, why? Was this an innate ability? Did I simply have better eyes than everyone else? Was my age something to do with it – are adult eyes better than a child's?

Had I eaten more carrots, or unwittingly prepared myself for this activity? At the time, I didn't really understand what had happened that disastrous night. Obviously, I had created more issues than I had started out to solve amongst my young subjects. But it wasn't until much later on, after I had been chatting to another naturalist, that I began to understand some of the science that had been going on in the background.

This early lesson in human sensory perception opened my eyes, in more ways than one, to the way we live and the expectations we have of our senses and, more importantly, how we can train ourselves to be better at using them.

Because we are very much visually oriented monkeys at heart (or head, as somewhere between 30 and 50 per cent of the brain's processing power is used for visual processing), it seems like a good place to start in our gradual 'rewilding' of ourselves. The first part of this statement very much holds true for all monkeys. It's a fact that eyes are the primary sensory organs for us and for all of the other apes and monkeys swinging on our particular limb of the phylogenetic tree. They're pretty big, too, relative to body size; our eyes, about an inch across and weighing in at a quarter of an ounce, are the largest. A fact we can probably thank the nocturnal habits of our earliest ancestors for.

As I'm sitting here writing, a 3.2-million-year-old news story has just broken. 'Lucy', the world's most famous fossil hominid, discovered by Donald Johanson in the Olduvai

Gorge in Tanzania, way back in the 1970s, probably died falling from a tree. Learning this made me think.

Getting a leap wrong would pretty quickly take that individual and all the genes they carried out of the gene pool while, conversely, those that were better at it, maybe with bigger, better and more forward-facing eyes, held a substantially better chance of surviving and, therefore, of breeding and passing on the winning traits.

Obviously, Lucy's was an unfortunate accident; she already had our eyes and some research suggests early hominids actually had better visual capabilities than modern humans, but as far as we know, she would have been able to reach out and grasp an object in front of her without fumbling around too much, just like you and I. Part of our excellent sense of sight has to do with our ancestral life in the trees and it's something we share with Lucy and many other primates.

One of the theories for our excellent vision relates to just this fact – our eyes are forward facing, giving us excellent binocular or stereoscopic vision. Stroll around a zoo or flick through an animal encyclopaedia and you can group animals into two broad categories, those that have eyes up front and those that have them placed in the sides of their heads.

It stands to reason that as eyes move slowly forward, the overlap in the field of vision each eye sees increases. It's that region of overlap where each eye sees the same scene from a slightly different angle of a few degrees that gives these animals, including us, something called binocular vision: the ability to judge distance and perceive depth accurately. The trade-off is that our overall field of view decreases as our binocular vision improves. In other words, we can judge distances better than, say, a rabbit, but we have only a field of view of barely 180° while a rabbit has around 360° and can see almost all the way around itself, in front as well as behind.

In 1922, a British ophthalmologist, Edward Collins, stated that early primates needed a visual system that would allow them to 'swing and spring with accuracy from bough to bough . . . to grasp food with their hands and convey it by them to their mouths'. As our ancestors took to the trees they needed to be able to escape their predators and capture fleeing prey; it makes total logical sense that evolutionary pressures on these early primates favoured a visual system with excellent depth perception.

This theory seems very plausible, until you bring up squirrels. Squirrels are just as capable of death-defying leaps as are monkeys, and yet they have eyes on the sides of their heads! To counteract the squirrel in the ointment of Collins's theory, another was born.

In 2005, Matt Cartmill, a biological anthropologist, proposed a different idea: predators need forward-facing eyes in order to pounce accurately on their prey. If you think of owls and cats, this makes sense. Early primates were probably insectivorous and, like today's bushbabies and tarsiers, would have hunted more by sight than by scent. They were predominantly nocturnal and this dependence on the eyes meant that in evolutionary time as the eyes and all the neurological plumbing that went with them moved forward, the physical space required for a nose and the various nervous connections needed to convey this information to the brain got squeezed out and thus one primary sense was swapped for another.

However, there are predators which don't fit this particular model either, mainly those that hunt by daylight, such as the mongoose.

Again, another honing of the theory was needed. Neurobiologist John Allman proposed that this theory could best be explained by predators that evolved to hunt at night. He reasoned that forward-facing eyes are much better at gathering light than those positioned at the side of the head. Our early ancestors probably operated at night and so this may well explain the way things are. A further neat addition to this was the 'X-ray vision theory' proposed by yet another neurobiologist, Mark Changizi. He suggested that our excellent stereoscopic vision comes about from our ancestral need to be able to see through dense foliage, hence the suggestion of superpowers in the name of this theory.

You can demonstrate this 'X-ray vision' yourself by simply holding a finger up in front of your face and looking beyond it – you'll notice two visual images of your single finger and both of them, due to the offset of your eyes, appear to be transparent. You can therefore see through your finger – yes, you have apparently got X-ray vision.

Superimpose this phenomenon on a different scene, one in which a timid, highly strung prosimian isn't holding a finger up in front of its eyes but is sitting cowering in dense foliage, attempting to stay hidden but at the same time surveying its three-dimensional world for prey, and this ability to see through the clutter of a thick and luxurious tropical forest starts to make sense.

Seeing the light

This all goes some way to explaining the positioning of our eyes, but what about the actual internal mechanics of the eye? In some ways this is even more relevant to what this book is all about; after all, there's not a lot any of us can do about the positioning of our eyes, but how we use and process the sensory information in our 'mind's eye' is sometimes probably more critical to 'rewilding' our heads than the raw materials we have at our disposal. This is something we'll explore in more detail in later chapters.

To truly understand ourselves, sometimes, just like with a machine, we need to take things apart, to deconstruct. For when we know what we are made of, we can play to our own strengths and work out a better way.

We now understand why and how come our eyes are on the front of our heads, and the advantages they give us by being so positioned. However, the really interesting stuff occurs when the light enters the eyes. The human eye is a remarkable thing. It has the ability to detect an extra-ordinary range of the light spectrum over nine orders of magnitude, ten million colours, and from dim starlight to bright daylight; however, it is limited to a much shorter range at any one time. How it achieves this is by an alchemy of light, proteins and electricity. It is so seamless and beautiful that it may as well be magic and, to understand this trick of light, we need to look behind the scenes and blow the conjuror's secrets.

As light passes through the cornea and hits the back of the eye's orb, several processes occur simultaneously. For a start, the amount of light hitting the retina triggers immediate adjustments of the physical eye itself. The eye is capable of controlling the amount of light that enters, by constricting the diaphragm of the iris, the coloured bit of your eye. This simply makes the hole, the window of the eye, bigger or smaller (see also pp.66–7).

The real magic happens at the back of the eye, the retina. This converts the energy of light into electrical impulses that communicate with the brain – in fact, technically, the retina is part of the brain, the only bit of the central nervous system that can be viewed without first using a scalpel.

The surface of the retina is comprised of two main kinds of cells, cones and rods. The way they work and don't work explains the way we experience the world around us, and understanding how they function the way they do is a massive help in squeezing the best performance out of your own peepers.

The first type, the cone cells, do colour. They evolved for day vision. They are, however, unable to respond to dim illumination. Just suppose you're gazing at a dazzling blue morpho butterfly, perched on a red flower, surrounded by the green foliage of the rainforest rather than the pages of this book. This colourful scene is colourful because the seven million cone cells in the back of your eye somehow tell you it is. Because we are animals with trichromatic vision we have an excellent ability to recognise and distinguish a range of colours in the visual spectrum, maybe as many as ten million. That is thanks to the fact that we possess three different flavours (hence the 'tri' in trichromatic) of cone cells. Each of these is sensitive to a different colour. They are blue, green and red or, to put it another way, different cones are sensitive to different wavelengths of light – short, medium and long wavelengths. The light bouncing off the morpho's wing is causing great photo-excitement in those cone cells sensitive to these wavelengths at the short end of the light spectrum and is therefore understood as blue, while, simultaneously, light bouncing off the flower's petals and the surrounding foliage is getting the cells in each of the other two types of cone cell worked up.

The other types of cells are the rods, 150 million of them, and they are over a thousand times more sensitive to incoming light. These are responsible for our night sight. To say these rod cells are sensitive is an understatement of the first order; they are capable of detecting a single photon, the smallest common denominator of light energy. However, during the day, they don't work – they're saturated, over-stimulated, pounded into submission by all the numerous photons bouncing around.

Colour inhabits the well-lit world; as the number of photons reaching us from the sun diminishes, then the cones stop being stimulated and those light-sensitive rods come into play. But because they can't differentiate between differing wavelengths of this weak light we therefore simply see in tones, which explains why we see in black and white at night.

The way cones and rods respond to light differently is responsible for several commonly experienced visual phenomena, such as that sensory miasma you get when you turn off a torch or step out into the night, something called 'dark adaptation'. In our over-illuminated world, this adaptation process is a vital component of our understanding of personal rewilding. As we learn to switch off from the lights and switch back to our natural senses.

First, one needs to understand the basics of what goes on at a molecular level in the cells themselves, the underlying chemistry of sight.

The only light-mediated event in vision is when a photon of light interacts with the opsins (light-sensitive proteins) and rhodopsins (a light-sensitive pigment) found in the rods. They contain a molecule called retinol. When the opsin or rhodopsin is hit with a photon, it changes to an active state, a process called photoisomerisation; shedding the retinol and setting off a complex chain of chemical reactions, it ultimately manifests itself as a neurotransmitter, which sends a nervous signal to the rest of the optic system and back to the visual cortex of the brain. Once it has released its retinol molecule into this chain, the opsin, as far as the visual system is concerned, is 'chemically down'; in this state it is known as a 'free opsin', and it needs to replenish itself and reset, in order to be on standby for the next photon. This replenishment of the free opsin with an untransformed retinol takes a little bit of time and it is this time which causes a delay in our optical system – the delay we consciously feel when we adjust to different environments. The interval that this takes to transpire is known as 'dark adaptation' and varies in speed between cone cells and rods. Cones are very fast to adapt to changes in illumination – think of coming in out of a bright sunlit street to the dingy recesses of a shop: it takes a little longer to adapt than if you left the shop to go out into the street. Rods, conversely, are much, much slower – a process which we all struggle with and one which, if we understand it better, can make us a much more visually effective animal.

4

Darkness Is Light Enough

W E SPEND over a third of our lives with our eyes shut as we sleep; and, logically, being diurnal creatures by nature, most of this shut-eye is taken during the hours of darkness. This means that by the simple act of changing your activity pattern, you can open up a whole new world of experiences, a parallel world superimposed over the topography of a landscape that is more familiar. It's a real frontier to be pushed at, an opportunity that should be irresistible to the inner adventurer; it's a time to push your own physiological boundaries as well as to conquer some intellectual and physical challenges. The potential for adventure is huge. Those eight hours a day are there for the taking.

A stroll at night, even in the built-up confines of the city, can open up a new, fresh world. Simply being on your own, away from the crushing confines of the rest of our species, can release us, so turn off your phone, tear yourself away from the mesmeric trappings of modern life, and step out into the welcoming, exhilarating blackness of the night. I say blackness, but it isn't really.

To experience the dark – I mean real, complete, utter nudity of light, where not a photon can find you – you have to be underground in a cave system or ten thousand metres below the ocean's surface in the Mariana Trench, and even here creatures exist that mess with your senses – oddball life forms, each owners of their own light-producing technologies.

To be underground or that deep underwater is not a normal place for a human; we are, quite literally, out of our depth. Such places are properly scary to me. I've been four kilometres underground in a cave system, with no lights; I've even dived in them. Trust me, and I'm not afraid to say it, these are places darker than your imagination. Even the fact that you might be imagining them now requires you to illuminate them in your mind's eye. I'm talking about total darkness, not a glimmer, not a single photon of hope. My fear of these places comes from a complete and utter reliance on our own cleverness, our own technology and life-support systems in order to be able to visually perceive them.

No, the dark I really want to explore is the accessible one that few of us deliberately go out of our way to experience – the night, the non-dark. It's a time in the daily cycle which, at best, we reluctantly tolerate, when we're walking home from the pub or dashing between the front door and the car.

Out there, beyond the arc of the streetlamp and the hypnotising artificial glow of seductive screen, one that is persistently pedalling counterfeit self-improvement and phoney escapism, lies a liminal, alternative world of new experiences. It's a calming walk into the wild of your inner turmoil. Disconnected from the source, these spectres of artifice fade away in the face of what is real. A walk that takes us beyond what is familiar and into the present in a way that is profound. It tests our senses.

Dance with your demons – the fear of the night

The rasping, dissonant scratch of its fur as it slid along the skirting board towards my bed woke me with a start. Quivering beneath the duvet, I could distinctly hear the hollow tap, tap, tapping of its curled nails on the painted wood draw closer. A metronomic counting of seconds before it got to me. It materialised out of the moon shadows, a spidery form, misshapen limbs scampering and lurching towards me . . . those bony fingers. Tap, tap tapping into my psyche.

As far as I know, I'm the only naturalist to get the night frights around one of the rarest and most threatened lemurs. I'll qualify that before you lose faith in me. I *was* afraid of aye-ayes.

It was probably a half-watched late-evening documentary while I was less than awake that started my irrational fear of this Madagascan rarity. My imagination would take hold and I would turn the rustle of a mouse in the loft space above my head, the shadows playing on my wall and the crumpled, discarded clothing on my bedroom floor into demons. The problem with a fear of the dark is that it isn't the dark we're afraid of at all; it's what the darkness may hide.

I have since spent a night alone in the Madagascan forest, I have faced my childhood nocturnal demon. I awoke to it pattering through the branches and to its eerie mechanical ticking and clicking as it percussively foraged, tapping the branches in search of the hollows and burrow chambers of insect grubs beneath, acoustically sounding them out, before winkling them out with that same peculiar, slimmer-than-a-pencil bony middle finger. Facing fear is what it's all about. Pretty quickly I realised that my aye-aye, although looking very much like a harbinger of all sorts of bad news, was a pretty eccentric and huffy, harrumphing character of the Madagascan night; an endearing and highly endangered unique take on an ecological niche normally taken by woodpeckers. Only these birds never got a zygodactyl toe-hold on Madagascar, so the lemurs got in there first.

An interesting aside to this is that while I was getting to know the aye-aye, I discovered that my childhood fear of this creature, while being completely irrational, was not something so odd to the Malagasy people. Who, it turned out, saw the aye-aye as an animal of ill-omen, a creature of the spirit world, one that was in their language *fady*.

It's fair to say that quite a lot of us are afraid of the dark, and with good reason. We're not really designed for it. We are, after all, visual creatures who need light for our eyes to work, and seeing is our primary sense. We need it to make sense of our immediate surroundings. When the light is turned off, or, since it is the sun we're talking about, when we're slowly rolled away from it, we become less functional. Sensorially deprived, understandably we start to feel vulnerable. This vulnerability was probably originally rooted in simple survival. At night, those creatures over which we would have held dominion by day suddenly become our retaliating masters. Many animals without the compromised visual perception we have, and with other super-senses, now would hold the whip hand. The hunter becomes the hunted, the table has turned. Crippled by the thickness of the night, we freeze, retreat, hide. This is fear of the dark at its most rational. In a paper published in 2011, one thousand lion attacks on humans between 1988 and 2009 were analysed in relation to the time of day the attacks occurred. The results are certainly enough to make you think twice about being active after dark: 60 per cent of these sometimes fatal encounters occurred between 6 and 9.45 p.m. If you live in Tanzania, you have good reason to be much more cautious after dark. So, despite evolutionarily climbing down from the trees and becoming clever primates, led more by our intellect than by our innate instincts, it seems that the second the light goes out, the candle is snuffed or the switch is flicked, some of us immediately revert to our primal state. Our wild self.

The irony isn't lost on me that now I'm going to encourage you to take a step into the dark and, for most reading this, that'll be into a land that is missing its big predators, which are, in many cases, the ultimate goal of rewilding. However, many of us still need the 'fear'; it's a cerebral requisite. We need to be scared of something so, while once this reaction was rational, in our modern world, for most of us at least, we demonise other creatures, people, spectres, and replace the real with the products of our imagination.

The best way to conquer this nyctophobia, as with any fear, is to work your way through it, to face the very thing that challenges you. You need to confront your aye-aye, whatever that may be. You'll find a walk at night is not only liberating, but opens up many other opportunities and benefits to you as a rewilding soul. I appreciate there is often a 'human element' to this fear of the dark. Many of us simply can't face the media-enhanced risk of meeting a less-than-savoury member of our own species. So with this in mind, it's probably best to start in an environment that feels 'safe' in this sense. You'll quickly find your enhanced perceptive abilities will empower you to own the night and eliminate many of these fears too.

Why it's good to go out at night

Our modern existence means that what was once a time of reflection, a time for the processing of the day's events, whether at school or at work, is no longer sacred. In our ever-switched-on life, an electronically generated 'ping' punches through all. Where once the nine to five meant that we could partition and separate our working lives, whether professional or educative, from our personal existence, now we are bombarded with electronic detritus. We inhabit a place of incessant communication; phones, no longer tethered to walls, follow us, a phone book of 'friends' you've never met stalks you from the shadows of your pocket. E-mails can get us at any hour, we live in constant fear of missing something; Facebooking, tweeting, 140-character digital epistles can sidetrack our thoughts and push us out of the moment any time we might try and have one. Our modern world effortlessly and rudely pushes into the dark hours from dusk to dawn. We can be inconveniently 'jumped' by our gadgets of convenience, ambushed into believing that a response is demanded right now. Ever desperate to get ahead, we live in fear of missing out, fear of abandonment. A new-world fear in the night?

Our digital persona encroaches on what was once a very personal time, a time of recuperation and reflection, a place for cognitive processing. That fact alone is a good reason to go out at night.

Being able to night walk, to step into the cool air, allows us to take back something that is rightfully ours. At its most fundamental, it's calming, a moment to decode and regain clarity, but it is also something empowering; it's a survival tactic. To face and overcome what you fear is one of the primeval thrills, a pragmatic matter of being able to deal with the demons in case you ever get caught out in the dark, something I've heard hill walkers term 'benightment'. But it also enriches our understanding of our habitats and environs. Even the most familiar daily constitutionals acquire another meaning when they are taken after dark, and to understand this and get a true sense of place is essential.

* * *

Recently, I reached the summit of Mount Kinabalu in Borneo and while it's probably one of the most accessible and highly visited of the world's big mountains, what made it so unique and special for me was the fact that I got to the top in the dark. Admittedly, I wasn't alone, I was with a motley assortment of ages all scrambling up the sometimes glassy-smooth granodiorite slopes to get to the summit before sun-up. It was noisy; there was a cacophony of clinking kit, rustling waterproofs and energy-bar wrappers as well as the talking, laughing, puffing and panting that goes with large herds of humans. Setting off from the rest house at around one o'clock in the morning seemed at one with the sacred mountain, although the head torches that were insisted upon for safety were not. Though I tried to avoid the glare of LED headlamps that nearly everyone was wearing, I was often left dazzled and, because of the pace of the climb, I found myself falling back on the necessity of using my own. In between the incessant blindings by those with poor headlamp etiquette, I did at least get to experience the celestially abundant sky, its clarity and contrast, the falling moonlight casting pools and saturating fissures, all unlike any experience I have had at home. And, of course, the whole point of leaving at that god-awful hour was to get to the summit, all 4,095 metres of it, to watch the sunrise over Borneo.

It reminded me of the joys of being out at night. You really don't have to go to the ends of the Earth or indeed the peaks of its mountains to experience the same magic. Experiencing the moonlight playing on a puddled rut, gleaming on the frosted rind of a winter's meadow or casting bewitching shadows behind a twisted thorn is a very rare and wondrous thing – it's there all the time, but it's just something few of us actually go out of our way to experience.

Try it and you will find it's like discovering a different side to a best friend. A walk through the moonshine and shadows makes even the most familiar and mundane magical. A little extra bonus is that in the temperate regions where winter days draw short, sometimes we feel a rush to get back to the hearth, to return at a pace, running ahead of the rapidly falling night, tripping over ourselves to avoid the enveloping dusk. Why? Walk confidently in the face of nightfall. If this is something new to you, here are a few basics to ease you in safely. If you are armed with knowledge and a bit of preparation, a night walk is one of the most thrilling, soulful and easy-to-achieve everyday adventures.

How to moonwalk

The earth has turned again, we've rolled away from day, the last light has faded in the west, things have got darker, and we've made the transition to the shadowy side of things. So let's get you out of the door. The first thing most of us who don't live in the perpetual daylight of the city do is grab a torch. This dependence on our technological salvation is what we touched on before. As a species we've managed to stretch the day into night, to invade the dark sanctuary with the arc of our artificial illuminations. Darkness is a modern inconvenience that we've learned to modify; we tackle night armed with Tungsten, Halogen and Fire, our own personal constellations and non-lunula light. There is a case for artificial light when exploring the night but first let's go native.

It may come as a surprise to you but at least on some nights we all have a degree of night vision, and while you'll have to give up on stimulating the colour-sensitive cones in your retina, you mustn't forget the special qualities of your rods. They come into play when high sensitivity is required.

Night-time is never a lacking-of-light time. For a start, there is the crepuscular light of dawn and dusk: on a clear night, the sun's fingers take a surprisingly long time to let go of the day and to fade to black, and even then it has never truly disappeared. There is always a degree of lunar luminosity to be had. Even if it is being reflected off the moon, the sun still has an influence of some kind, albeit minimal on a new moon. When there is a full moon, that silvery disk is still reflecting 7 per cent of the sun's light down to us, a light that mainly peaks at the blue end of the spectrum. This provides us visual monkeys with plenty to work with. Eyes can still perceive this light, even if it isn't flooding our visual cortex with information in the same way as it does in the day-time. Night vision sounds like another fanciful super-human quality, yet we all have the ability to pull back that velveteen veil that descends on the last curtain call of dusk.

The iris of the human eye, the coloured bit, works like the diaphragm of a camera and is controlled by paired muscles.

These cause the iris to contract or dilate, and this in turn controls the amount of light getting through to the retina. So when the light is bright, the iris contracts and the pupil, the window through which the light has to pass, gets smaller. However, at night the eye needs as much light as it can get; your pupillary window is thrown wide open, which, in a young person (the ability and strength of the muscles that do this wears off over time), means a diameter of seven millimetres. I say 'thrown' – this is misleading; it suggests an instantaneous reaction, and you will shortly see why it is relevant to your night vision. It takes about fifteen minutes for a pupil to dilate to its maximum aperture. That's fifteen minutes after sunset – or, indeed, any other light stimulus, be it torch, headlight, streetlight or lamp – for your eye to let in its maximum amount of light, and another half an hour or more for your retinal cells to adjust. Remember the detailed explanation where I blinded you with science? That stuff about photoisomerisation, retinal and 'dark adaptation'? Well, this is where it becomes relevant: that half-hour or so is the amount of time the free rhodopsins in your rods take to reset themselves, to recombine with unaltered retinol. Your cones are resetting too and, much faster, but because they will not be turned on by the low light levels you're experiencing at night, they're as good as useless.

So, in total, that's around forty-five minutes – three quarters of an hour in order for your eye to work at its most efficient at low-light levels.

* * *

The first step to better night vision and to releasing your full potential is to starve your eye of artificial light. My favourite way to do this is to simply ride the dusk, to allow the night to fall on me. This is what is natural, this is what we evolved to do. In nature, lights rarely just come on. The slow and gradual switch from cone-led coloured vision to black-and-white tonal sensitivity of the rods is a gentle and gradual affair, a secret transition to night mode. This feathering into night is what I had done, unwittingly, prior to meeting up with my young night-walking disciples mentioned at the beginning of the previous chapter. The problem was that they had not.

The same issue applies if you are going out later, whether you've been driving, been indoors, or, indeed, have looked into a torch. Just a moment of exposure to any source of bright light and your eye is reset to default, your cones are stimulated and respond fast, the iris shuts down, the rods are swamped and your night vision is once again temporarily compromised. Night vision is therefore a fragile thing, slow to develop and yet speedily lost.

The reason the majority of us are not confident at striding out into the night is that few of us have even experienced that full nocturnal capability of our eyes.

It's a sensory handicap we probably unwittingly have given ourselves, from the time we first huddled around a fire to escape and diffract the terrors of the night. Since the moment we discovered fire, humans have been stranded, isolated from the night, swimming around nervously in their own pool of light, peering into the periphery of the penumbra of the darkness – which, if we were to pluck up the courage to turn that light off or blow it out – would reveal a world that would seem a little less scary.

There is no short cut. However you might want to experience the night, you have to give yourself a period to condition your eyes. This is something I learned, partly by chance, the hard way.

One lesson that sticks with me is a fairly spectacular accident that involved my heading out into the woods to watch a family of foxes with cubs. Setting off for my high seat in a rush, I wanted to make the most of the dark hours before my curfew; it was a school night and I had to be back at home before it was too late. Without wishing to seem a bit long-in-the-tooth, these were the days before efficient batteries or, indeed, bright, portable lights such as we're blessed with today. I got to my high seat, a home-made affair, an old rotten wooden Windsor chair pilfered from the local dump. I had sawn its legs off and tied the remaining portion with homemade knots to a tree branch.

The whole crude and rustic construction was precarious to say the least, but when I was seated in it, I could sit some six metres above, and out of the wind of this rather boisterous family of foxes. The only challenging bit was not the climb up the tree, I could do that with my eyes shut – a series of easy horizontal boughs and a natural lattice of ivy, with perfect footholds and handholds – it was the jump and twist required to get from facing the tree trunk to flipping around and facing the other direction, while landing in the delicate homemade high seat. I had done this dozens of time to date fairly uneventfully. This night I got into position in good time and without incident. However, I had forgotten to check the charge on my torch before I left the house. By the time the fox family had emerged, scratched, played and performed their querulous antics and then had got bored and had slunk off to find mischief elsewhere, it was dark. My time was up, and now I had to get down. I flicked the switch, and just before I hung the torch around my neck to perform the tricky flip manoeuvre in reverse, there was a gentle *tink* and the light went out. I desperately flicked the switch back and forth with my thumb, trying to coax a little more life from the battery. I only needed a few seconds, the rest of the ten-minute journey home I could probably manage at a stumble; I had done it so many times that it had become first nature.

However, I had no light and just as little choice. I had to try for the tricky gymnastic manoeuvre not only in the dark but now also while I was night-blind. I had one go at this and I blew it. A bumbled footing, a missed handhold, and I was hurtling towards the ground. I plunged to the woodland floor, painfully making contact with all the branches that had been my friends on the way up. Now each of them sucker-punched me as I bounced off them, each leaving me with a bruise that would turn purple by the morning. I hit the ground with a thud, having first passed through a bramble thicket. I ended up flat on my back, gasping for a breath that seemed to take an eternity to come. A school of hard knocks indeed.

At about the same time as I was tumbling out of trees, I got my hands on a copy of a book that inspired me to make friends in the night. A magical book by Chris Ferris called *Darkness Is Light Enough* (from which I borrowed this chapter title). It is the story of an inspiring woman, plagued by back pain and insomnia, who walks the woods and fields around her home in the Scottish borders at night. It's an enthralling account of a naturalist and the nocturnal wildlife she shares the night with. The book is packed with information on animal behaviour and the less-than-savoury intentions of poachers and those intent on illegal badger baiting. It had me hooked, it rang true, and in so many ways her observations had mirrored many of mine – but the bit I couldn't really get my head around was how she describes walking around without a torch, following her animals as they went about their nightly business. Up until that point I had only ever been stationary in my night-time watch. I had a torch with me, and if I ever did stumble across a badger, fox or deer while I was out after dark, on nearly every occasion up until that point in my life, the animal fled, in a thudding, crashing sort of panic, the sort that creates as much distress in the human protagonist as it does in the animal. It was as if the woman had super-human powers, something I thought I could never aspire too, mainly because I imagined them to be artistic licence, untruths used to aid the otherwise believable narrative. That was until I tried it myself.

I would walk to the bottom of our long, narrow garden, far from the warm glow of the kitchen window, and wait for the security light back at the house to go off, then I would turn off my torch. I would be completely blind. To keep walking was self-inflicted madness; I would often stand, paralysed, as soon as I reached the limit of the ambient light that issued forth from the lit windows of the house or the moment the torch beam was extinguished. But not blessed with patience and without the knowledge of what I know now, I would try and force things along.

I got tangled in fences, fell in the old bath tub (sunk in the ground and used as a pond by our domesticated ducks) and stank of anoxic pond water for days, hit my head on branches and committed myself to spectacular trips, stumbles and unintentional roly-polys down embankments. It wasn't going well.

There is no rushing things, which is probably why the majority of us don't see the night through the right kind of eyes, through our unfettered wild eyes, that is. We live in a world of no temporal space, with no time to catch our breath. Frenetic rushing and racing, never a minute wasted, never a minute to spare. There is no instant gratification when it comes to realising your nocturnal potential. Can you imagine not looking at a screen of any kind for forty-five minutes? That means not a call, not a tweet. Most would have a technological meltdown. Then there is the other conundrum: what can you do in those minutes? That's why it's easier, if you're a beginner, to simply go out for a walk while there is some daylight and then you can naturally slip into night-time mode. Otherwise, you can meditate, prac- tise some mindfulness, or even do some stretching and warming up; the more supple you are, the less likely you are to injure yourself if you do stumble or fall.

Fifty thousand shades of grey

To make it even easier, certainly if you're new to night walking, choose a night with a full moon. Assuming you've got a clear sky or at least only a thin veil of cloud, it is in these conditions that you will have the maximum 100 per cent reflectance of the moon's shine that'll keep you company from as soon as the sun fades at dusk to when it reappears at dawn. Tuning in to the moon's phases like this is an awareness that once was vitally important to pre-domesticated human culture. A full moon simply means that it is in line with the sun – with us, Earth, in the centre. Therefore without our obscuring shadow, it shows its full face to us. It was a welcome one to our predecessors; it was a clear point of temporal reference. Compared with the sun, it marked the passing of time with clarity and a night full of moonshine was always a special one. Our modern almanacs are still based on these movements of celestial bodies in relation to each other, our months or 'moonths' originate from watching and counting the appearance of celestial bodies in the sky.

Before effective modern lighting was available, a full moon would have provided an extension of our diurnal activities. In the Northern Hemisphere the moon's complete disc is given descriptive names that reflect its importance by its various nature cultures, nominal notions of the seasons with a far greater meaning and resolution than a mere monthly moniker. Descriptions such as hunger moon, snow moon, moon when geese come home, hunter moon, sap moon, planting and beaver moon, all illustrate a practical and, in many places, previous significance. To know the moon is to really know where you stand in the calendar, rather than relying on the cold and the digital, a knowing of the moon is a vital part of understanding the world, and gives those open to rekindling this relationship a solid step up the ladder of rediscovery. To know it so is not born of a sentimentality or nostalgia. It's a cycle, around which much of the natural world is attuned, and therefore to discover a sense of place, you need first to become familiar with a true sense of time.

For your first moonwalk, it is essential to have a full moon and calm conditions, with little or no wind, in order to build your confidence. Some would even go as far as to say that good conditions are better than a good moon. A quiet night means that your other senses, which you'll start to use to a higher degree too, now that you're walking on the dark side, will not be compromised. There is nothing like a stiff wind blowing to set you on edge, and heighten tension. The incessant agitation of the flexible. The knocking together of countless leaves and stems, the rattle and scratch of innumerable clave and gong, this is the timpani of nature, and is of course part of the music, but it also competes with other clues as to your whereabouts, and this disassociation with such a critical sense and the feeling of anxiety that naturally comes with it possibly again relates to a deeply buried rational survival mechanism. The footfall of a predator cannot be heard. Even now the sudden, whirring flight of a startled woodcock as it springs from a woodland path, croaking into the night, makes me jump out of my skin more on a breezy night than on a quiet one; maybe I can hear the more subtle precursors of the bird tensing up for flight? This nervousness is also shared by many other night walkers – nocturnal wildlife, birds and mammals are more jumpy and nervously predisposed too. It makes for a much more difficult task to watch anything, from rhino at a watering hole to hedgehogs in the herbaceous border, if they're on high alert.

Walk where there is a clear view of the sky; the closest you can get to pitch-blackness is when you're under a leafy canopy or moving through a coniferous forest. Another enjoyable way to wait out the period of dark adaptation, to adjust and find that magical state, is to simply watch the sky. Look out for the first celestial lights to show – this will usually be a planet and may be in the sky opposite the setting sun, as here there will be less interference caused by the sun's remaining rays bouncing about. This familiarity with the night sky then takes on another dimension. Become familiar with the planets and constellations and you will engage with the world in a way that was familiar to early navigators. When you're advanced, this will serve as your compass and aid in your navigation while night walking.

Get the perfect conditions and walking under the moon becomes a magical experience, almost super-natural.

It doesn't seem right, but you can see almost as well as by day. Your rods are revelling in the relucent qualities of night; they're filling your visual cortex not with just black and white, but with, some say, up to fifty thousand shades of grey.

Where you walk is worth bearing in mind too. Certainly on your first few attempts walk in places that you know, choose landscapes that are familiar to you. Even the most un-stimulating twenty-minute constitutional takes on a new lease of life when carried out under the veil of night, so don't get too ambitious on your first attempt. Ignoring my own advice, I've managed to end up on the edge of a small quarry by a cliff, hidden by a dense stand of bracken on a slope that I had thought I was as familiar with as the back of my own hand. Let my walk home, covered with purple bruises and bramble pricks after that oversight, be your lesson.

Find yourself a location where there are few physical obstacles – one without too much in the way of challenging topography and trip hazards is the perfect classroom. In short, don't take too big a section of the night; keep it really simple. Choose places where the paths are big and wide and also, ideally, somewhere not too far from roads or other access, just in case you feel the need to bail out at any point in the venture.

Reflective water can be a boon to the nocturnal wander. The surface on a still night shows a bright reflection of the sky. Moonlight bouncing from saturated potholes, ruts and puddles illuminates like the emergency floor lights in an aeroplane showing you the path. Larger bodies of water, lakes and rivers not only give you some clear boundaries at night, but they also bounce more light back into the surroundings.

The night can disorientate you and if you're not used to it, you can get lost in time and space very easily, so, in addition to choosing familiar ground, certainly at first, learn to navigate – and I don't mean by Google maps. Handy though the smartphone is, its batteries don't last forever. I've had a GPS system running in my phone, keeping a track on my nightly ambles before. It's fine, but it tends to eat up battery life fast – a battery that you might need for an emergency call. You can, I know, take a booster pack with you, but then you need to carry that, plus the cables to make the connection, and, before you know it, you're lumbered with stuff. The technological burden you're trying to get away from now has you laden like a stumbling camel. No, when I mean you should learn to navigate, I mean in the tried and tested analogue way. If you're truly into self-wilding, then I recommend you go on a natural navigation course; the environment holds plenty of clues as to where you are on the compass, and there are lots of folk who teach this old but essential off-grid lesson. A compromise between the two is a map and a compass – but don't just pack them. Learning to use these simple and beautiful tools is in itself a useful life-skill.

For me it's a time of solitude. Night walking is a time of personal reflection, when I can let my consciousness expand into the limitless night. I can let my senses heighten and I can settle into the blanket of the night, attuned to its rhythms.

An important part of the experience can pop like a bubble when the person next to you decides it's time to unwrap and eat a Murray mint they've just found in the fluff at the bottom of their hand-warming pocket. People are lovely but they do make a lot of noise, even if they're trying to be quiet.

What you get out of a night walk or nature in general is often something very personal to yourself. Nobody walks, behaves, senses or interprets the same thing. It's what makes it so special in the most basic sense of the word.

We'll cover this in more detail later, but if you're on your own, you can concentrate on the night, you're not followed by the background rustle and clump of another human, you don't feel any kind of obligation to stop, start, look or not look, and you're not plagued by that social human impulse to talk. Having said all this, sometimes for your first go at it, or if you're still at an age when your parents would rather you had company, then accompaniment can help a bit. What this does, however, is to detract from the main goal, which is a rewilding of yourself. If you take someone else with you, you do, to some degree, become a little less empowered by the night; you miss out on the thrill, the buzz, which is the coming to terms with some of your primal fears, fighting them out of your head and being able to experience night at its most pure.

Think back to what you've learned about the mechanics of your eyes. Because the focal point of your eye, the *fovea* is crammed with cones which only work by daylight, looking directly at an object, be it animal, plant or landscape feature, isn't much use. The act, which is born of muscle memory or instinct, of looking directly at what interests you, needs to be relearned at night.

The highly sensitive rod cells surrounding the fovea are the vital components you need to use at night, which means you have to look off-centre. Is that a wallaby or a wombat? Look to its right or left and you'll be able to tell, as you are now training the image onto your rods. While the resolution and therefore the detail will not be as good as a similar experience in daylight, because rods don't do resolution very well, you will at least see something, which is more than can be said for relying only on your cone cells, as they haven't even been roused into action by the moonlit scene. Something you may also notice when looking at subjects more on the periphery of your vision is that your rods are much more sensitive to movement. Often it is a flick of a wing or a nod of a nose that you pick up on in your periphery view which leads to you noticing that roosting robin or the giant jumping rat scenting the air at its burrow entrance.

If you're going to need artificial light at all, and sometimes you will be grateful for it, use red light. As we've seen, being kind to your rods is the best way to see in the night. To understand how they work, and the fact that they simply didn't evolve to work in a world where a headlamp or smartphone screen would suddenly flash at them in the dark, helps us to get the best out of them. To know that their light response peaks in the region of blue light is also useful; after all, these rods are designed to work under nature's night lights. The moon's reflection of the sun is not silver, it's actually blue – our eyes are naturally tuned to the blue moon.

This effect can be seen in some interesting phenomena. A red rose, of course, looks red under daylight conditions, those colour-sensitive cones rule, they predominate, they allow the rose to pop out at us against the somewhat more subdued green of the leaves in the background. But as the light fades at dusk, the cones start to shut down, no longer stimulated by the lower levels of luminosity, and the rods take over. The rods are not stimulated by the long wavelengths of light from the red flower, but before they switch over completely to night vision the much shorter wavelengths of the green leaves predominate and there is a period where the green appears much brighter. It's a brilliant demonstration of how the eye works and a fun way to while away that forty-five minutes at dusk.

This human insensitivity to red light is also the reason aircraft cockpits and some modern cars have red dashboard illumination and why some pilots wear rose-tinted spectacles – it reduces the risk of being dazzled and left in the dark. An added bonus for the nocturnal naturalist is that other creatures are not very sensitive to red light either, it's practically invisible to them.

Returning to the back of the eye: nocturnal animals operate on their light-sensitive rods, just like you do with your torch switched off. They operate at a peak spectral sensitivity somewhere in the region of a wavelength of 505 nanometres (the cyan-blue of moonlight), meaning that anything longer, such as a deep-red light of around 650 nanometres, is invisible to them. However, because we have those colour-sensitive cones, we, conversely, are able to see this light and therefore perceive detail, and thus are able to watch without being seen.

A trick I used to perform before the days of the Internet made sourcing materials so easy, and before the evolution in torch technology, was to collect red sweet wrappers, or red cellophane – a rare commodity in packaging back then – and stick them together to form a filter big enough to cover the end of my torch. This had the effect of creating a red light. But it was an unsatisfactory one due to the poor optical quality combined with the low power of my torch. The whole affair made for such a pathetically dim glow that while it probably didn't disturb nocturnal animals visually, there was always the risk of the excruciating sound of crinkly sweet wrappers if I wasn't careful. More often than not, I was reduced to watching creatures I could crawl right up close to, those that didn't have ears – snails and beetles rather than deer and badgers!

Nowadays you can purchase torches with built-in red filters for just this purpose, and modern LED technology can push out so many lumens that even with a red filter, the light can be thrown far enough to see more distant subjects. So, if you start to get serious and can justify the financial outlay, buying a good modern torch with a red-light filter is well worth hours of stumbling around blind or standing around waiting to become acclimatised to your surroundings again.

Even on the most illuminated of nights, when a distant woodland seems as three-dimensional as it does by the light of the sun, the dark veins of hedgerows are traced over the silver fields, and night creatures as small as a mouse can be viewed as long as they shy away from the shadows, the details and textures will still be missing. This unfortunate lack of perception occurs because the laws of physics have reached the limit of your lucidity; your rods are great at finding the little light that falls on them, but they're no match for the tightly crammed, super-high resolution of the cones that are stimulated by daylight, the sort of visual resolution that a diurnal primate has become accustomed to. So a student of the night has to decide on the experience he or she wishes to gain from the venture. The micro-magic of the smaller inhabitants of the night are as much part of it as the moon itself, yet to those intent on perceiving the night as a natural experience, untainted by electronic lumens, they will remain invisible, unless you blind yourself to the moon and the stars.

Take the ethereal glow of the female glow-worm – she glows a yellow-green light from deep in the anarchy of the grass and briar and she's an embodiment of the romantic notion of a night walk; we all love to see anything generating its own light, from fungus to millipede to beetle, bioluminescence is such a rare and delicate thing that it requires completely settled and fully adjusted night eyes to see it properly. Yet due to our own internal physics and chemistry we are denied experiencing, in the way we're used to, the whole. Can we be satisfied with only the mystical, cool light generated from within the beetle's abdominal tissue, knowing there's so much more to see? After all, the natural light is only a small part of the story of our understanding of what is unfolding deep in the tangle of the path verge. I want to experience and possess this light, own this knowledge some more. I want to experience what the bug-eyed winged male might feel when finally he homes in on her quilted, egg-heavy body, maybe even get a double-whammy of luck and see him as well, to catch him out, already lured into the love-light?

So I resort to default and turn on my torch – it's killed the magic at one level, but sometimes it's worth it. Now I can understand the topography of the light source, I can see her very un-beetle-like form, a gentle curled 'C' clinging on to a grass stem with six stiff, spiky legs, the muted fleshy pink of her flanks soft against the hard integument of her dorsal. The glow of a glow-worm is only one side of the coin; the other is actually to see that it's not a worm, to understand this weird beetle, to really get to know it as a creature and to recognise it, when it's out of the context of the night, when you're flipping a stone or when it's out wandering. This is, of course, just as valid and lends another level to your experience and understanding, part of the ongoing accum-ulation of your knowledge of the night. I always carry a torch and, while I try to resist the temptation to turn it on, it is there in case of emergencies or in case of glow-worms and other minutiae that I might otherwise miss.

Another technological advancement that aids our under-standing of the night's inhabitants is the night-scope or image intensifier. I love them and loathe them. As a daydreaming (or nightdreaming) kid, during my pre-badger-watching days, I remember fantasising about being able to see at night, as if it were day. I wanted more than anything to see wildlife and the bigger and furrier the creatures, the more elusive they seemed. But just imagine what you could see if your eyes could penetrate all the intimate details of the night as an owl can – I could watch the foxes floating along field edges and the bumbling silhouettes of badgers would be so much more rich in texture, pinging with detail.

All this was complete fantasy – until, that is, my university days, when my mammalogy lecturer let on, in a moment of foolishness, that he had in his procession an ex-military night-scope. After much begging and pleading he succumbed, whether to fuel my enthusiasm or to simply shut me up I couldn't tell. The long and the short of it was that that night I headed out with his treasured bit of Russian technology. It was about eighty centimetres long with a diameter at the larger end of a good thirty centimetres. It was a beast: having to be carried over the shoulder on a strap, it looked every bit the piece of military hardware it was – the impressive heft was further compounded by the fact that it had a large magnifying screen which you could screw on to the eyepiece.

Dressed in a long black overcoat and black, brimmed hat, and with what looked like a bazooka slung over my shoulder, I set off into the woods behind campus and hunkered down not far from a badger sett. The first badger emerged. My pulse was up, I chose my moment when I guessed, by the scratching noises, that several of the animals were up, and then *click*, I flicked the switch. But no matter how slowly I performed this act, a noise was made: a definite well-engineered Russian *click*, but it proved no more than a rude interruption causing several striped heads to freeze. The very high-pitched whine of the unit then made the heads bob a bit, and although there was a momentary, reasonably good view of a group of greenish badgers staring down the end of the sacred scope, it was short-lived; they were off, freaked out by the alien experience I had just unleashed on their evening.

I had a little more success when I bumped into animals out foraging; they were a little more relaxed, and, also, there were other noises around campus which drowned out the continuous whistle of the light-gathering machine. However, the same couldn't be said for the amorous couple, students who had driven up to a quiet part of campus after the clubs closed to do what young folk (other than those running around with Russian night-scopes) do on a Friday night. The appearance of a man in black, curious about the noises of the wild night, with a face suddenly illuminated by a soft greenish glow, was not something they expected to see. They departed as fast as the badgers did earlier, crashing through the undergrowth. The night of mixed highs and lows had one last surprise for me: I was stopped by a patrolling police car on my way back home. This was my introduction to night-scopes.

Night-scopes are much better now. The newer genera-tions have much clearer, cleaner image, with less fuzziness; they're also quieter on the whole and many are much smaller, with some being compact enough to fit in the pocket.

However, the biggest downside is that, like torches, they will give you night blindness. The image down the eyepiece has the same effect on your retinas as looking full-on at the lens of an illuminated torch. If you're already set on using a torch, then maybe they have a place in your kitbag. But for me, given the expense of these toys, a much richer, natural, less intrusive and cheaper experience can be had by simply using a torch.

The pessimistic Boy Scout in me says plan for the worst. Put together a small grab bag of essentials – a phone, map, whistle, ID, compass and a walker's first-aid kit are important elements and can easily be carried in a small backpack or in your pockets. Food supplies and liquid are a good idea too (you can get surprisingly hot and sweaty climbing hills after dark); it's not natural to think about dehydration at night, but if you're out for a long one, then it's worth taking this into account.

No matter whether you intend to use it or not, always take a torch along with you; you just never know, and even if you don't use it, a torch can be a useful signalling device if, for example, you take a tumble and break or sprain some-thing. While I recommend going solo on nocturnal trips, it's always a good idea to let someone know what you're up to. I know it's not a sense we're covering much in this book, but common sense is a useful addition to your repertoire.

5

Learning to See, Not Just Look

'IT'S RIGHT there, use your eyes, open them up.' Mamy, my Malagasy guide, was beginning to sound desperate, as a sense of building frustration snagged on those last words. His pride was at stake here if he failed to get me to see what was clearly blindingly obvious to him.

'Follow the main trunk down, see the second branch you come to, it's to the left of that. You can see his eye clearly.' I couldn't. In my defence, I was trying to pick out the 'king of crypsis' – a leaf-tailed gecko, an animal that seems to be almost as much tree as it is lizard.

Not only does it resemble closely the tree it rests on both in colour and tonality, but its fine scaled body is decorated with such an accurate garniture of colours as to create an almost photorealistic effect. Lichen only just does lichen better than a leaf-tailed gecko.

To top this, nature's masterclass in camouflage, the body is squished flat, like a lizard that's been worked over with a pastry roller. In fact, every part of this lizard, from toes to tail, seems to be a handful of millimetres thick. A frippery of fringing to break up the outline even more, even a transparent eye-lid complete with a peach-stone filigree, a reticulation which means that even that dead-cert point of reference is well hidden. You can't see it, but the leaf-tailed gecko always sees you. It's a phantom of a reptile, a membranous slither of lizard life invisible to all in the Madagascan forest – all, that is, except Mamy.

What was getting Mamy so exasperated was that the gecko was so clear to him that it might as well have been painted Day-Glo orange. A fact further emphasised by the way in which he spotted it in the first place while driving past. But no matter how I stared at it, using all of what I thought were pretty good wildlife-spotting skills, I simply couldn't make a lizard out of the textured trunk. What made this even worse for both of us was that the lizard was also very close.

So close that we hadn't even got out of the car; the small leggy sapling of a tree sat on the edge of the laterite road, and I could have wound down the window and reached out and touched it. I'll be honest: at this point in time that was something I was very much tempted to do. You see, a leaf-tailed gecko's next mode of defence is arguably as spectacularly shocking as its camouflage is effective.

Touch one and what was tree bark a split second before magically pops up; a gargoyle materialises in a flinch. It opens its flip-top, fringed, almost crocodile-like head and screams like a banshee, while displaying the scarlet interior of its mouth.

A display that is so shocking it's made me drop my torch on more than one occasion before. My pride wouldn't let me do it, however. Even getting out of the car, and changing my angle, defocussing my eyes a bit, like you do with those equally infuriating magic-eye pictures, didn't work; and then, with the tree between me and Mamy, who was still in the car and holding his head, a small red-winged dragonfly briefly landed. There was a minuscular twitch of the bark, and suddenly my eyes and my brain seemed to click into agreement. A toe, unmistakable with its flattened distal tip, briefly curled up a tiny bit, then nervously flattened again and resumed its optical illusion. That was all the point of reference I needed; I could now slowly trace the shape of its outline on the tree. I was drawing by numbers, however, missing some of the really clever bits but linking enough of the lizard's perimeter for the illusion to dissolve, and the unmistakable outline of the lizard to materialise. A second or two and I had it, finally.

This first 'leafy' of the trip was difficult enough; I earned my stripes, though, even if Mamy nearly had a breakdown while I was going through the process.

Much of my professional life has been about beating nature at its game, challenging those rules and strategies that have taken millions of years to perfect. Some of my employment has been deploying skills that nobody in their wildest imagination would have thought could deliver a pay packet later on in my life.

As a caterpillar hunter, I spent weeks, day in and day out, bent double or prostrate, literally crawling around on my hands and knees, scouring the crispy, dry leaf-litter of last year's bracken for a small caterpillar. This insect, the larval stage of a rare and threatened butterfly here in the UK, the high brown fritillary, is no more than a few centimetres long and it's not only of the same auburn, black and grey mottling as its immediate habitat, but it's also decorated with a texture and patterning that makes it look exactly the same as a pinnae of the dead ferns on which it basked in the spring sunshine.

An entomological 'pin in a haystack'. My job was to locate them and, when I had, I was required to take various measurements of the micro-habitat in which they resided.

I was also instructed to tell my boss, by phone, when I had found them. For the first few days, I admit it now, I didn't find any; not a single one.

My inauguration as a caterpillar hunter was far from what I expected. I thought I was good at this kind of thing – after all, that is what I told my new employers when I first applied for the job. Now I was in a quandary; my pride was severely dented, my caterpillar hunter's ego was deflated.

What did I do? I lied. For the first few days, I told my boss by phone that all was well, caterpillars, although few and far between, were being found, but the neat boxes on my data sheet remained unfilled.

I knew I had placed myself in a predicament; failure to find any of the larvae was not now an option. I just knew that they were here. They had to be. I was not going to be beaten, and even if I had to turn every brown crinkled leaf to find them, find them I would.

I was reminded of Dian Fossey, when she first arrived on her study site and there was not a gorilla to be found. It wasn't until the third day, after two hours of searching, that I found one. I nearly burst with the relief of it. This perfect little thing, honed by evolution to play the trick on the eyes of anyone or anything that wanted to see it. But I had cracked it. Now I had found one, others rapidly followed. It was as if someone had unblocked a pipe and removed a constriction on my optic nerve; the caterpillars of the high brown fritillary now seemed to be etched on both retinas. It was never hard to find them again that season, although I did have to relearn the process in subsequent years after effectively taking ten months off.

I've had other analogous quests for hyper-cryptic species, including great green grasshoppers, stick and leaf insects, earth-star fungus, the minuscule eggs of a tiny brown hairstreak butterfly – even seahorses. No matter what the species or the kingdom to which it belongs, I've found that the strategy is essentially the same.

Fundamentally, it's the process of being a predator. While I personally have never consumed a caterpillar, satiated myself with a seahorse, chowed down on a gecko or poached a butterfly egg, the fact is, I'm simply deploying what behavioural ecologists call an optimum foraging strategy – not for actual sustenance, but the function is the same; the job at hand feeds me indirectly, it puts bread on my table and milk in my fridge.

The first lizard, insect, fish etc. takes a while, but when you've tuned in to your quarry's size, colour and texture and, to a lesser extent, the more subtle nuances of the sorts of places they like to hang out, then the discovery rate increases. My Madagascan trip, while a little slow to start with, was a good example of this. Before long, I was finding my own leaf-tailed geckos, though I never had Mamy's success – possibly because I was less reliant on my own abilities, for if I failed, Mamy was sure to succeed.

But as a solo, lone caterpillar hunter, I got spookily good at finding these canny creatures. The perceptual tool that helped me is exactly what an actual caterpillar-eating species would use – something called a 'search image'. The concept of a search image is that, when repeatedly presented with the same visual puzzle, you can learn to solve it. Famously, this was demonstrated with blue jays trained to peck at pictures of moths: if the jay pecked at an image with a moth on it, it got a reward; if there was no moth, it got nothing and had to wait some time for the system to reset and the experiment to continue. However, if the pictures of the cryptic moths were of random, different species, the jay's hit rate wasn't so effective, but if it was shown the same species, it got increasingly better. We are capable of the same. A set of visual cues are collected up with experience and seared into the mind's eye; a visual pattern, a shape that seems consistent in a chaotic environment – you're subconsciously tuning in to some really subtle cues, your high-resolution retinas and your processing power has found a friend, a short cut. These search images are very useful but you need to invest time in learning them, and you'll also find that they don't stick around unless you're using them, which is why, on another caterpillar season or trip to Madagascar, I will have to go through the whole frustrating and infuriating process all over again in order to crack these anti-predatory adaptions.

The same light is bouncing off the cryptic animal and into your eyes, whether you see it or not. Mamy's eyes were receiving exactly the same wavelengths of light, the very same colours as mine – they had to be; my head was right next to his, but something, born of his previous experiences and his accumulated knowledge of where to look and what patterns and search images to look for, gave him the foraging edge. He was in nature and by her rules a more successful predator than me.

It's clear that looking and seeing are very, very different things. When we are immersing ourselves in nature, we need to be fully aware of what is around us. We need to be open to everything, every clue, every nuance of the environment. For a naturalist this takes a bit of time to develop. Rather than a recipe which clearly shows the way, this is more about superimposing experiences one on top of the other, to provide a deeper knowledge. This is an accumulative natural wisdom, a true understanding, customised by your own desires and drives. In short, learning to see is not something you get out of books.

However, there are a few tricks and exercises that will help you. But first, I want to draw your attention to the following which will help you understand that we can always be more aware.

A common belief is that people deprived of one sense can develop the others to a higher degree, and there is some evidence that a degree of replumbing of neural pathways can occur utilising the processing powers of other parts of the brain, but the science and the understating of this is complicated and beyond the scope of this book.

A study by Gallaudet University in Washington, DC came to the conclusion that people with impaired hearing probably cannot see any better than those with full sensory capabilities, but what they do is see differently.

In the complicated visual world in which we live, we are bombarded with stimuli. We stand in the centre of a three-dimensional world; there is information about where we are and what is going on literally anywhere we look, and we cannot process all this information at once, so we have to decide what to give our attention to, at the expense of something else. How we prioritise this incoming visual information about the world, and therefore how and what we perceive in that world, is down to us as individuals.

An eye is an eye, after all. There may be small differences in how it's built from human to human, but, like a camera, the laws of physics work the same from individual to individual, from you to me. What is different between someone who notices that there's a moth resting on that tree trunk or that there is a wren preening beside the path and someone who doesn't is known as 'visual attention'. To continue the camera analogy, assuming it's a digital one, it's how we process that information, what the camera or the image-processing software on the computer does with the information we've given it that makes the difference.

If you have a full complement of senses, then you literally have 'all-round' perception; your other senses, particularly your hearing, help the eyes out. Think of yourself as the centre point in a sphere. Whether a sound or a visual stimulus grabs your attention, with a quick movement of body, head or a flick of an eye, you can be 'on it'. If you are deaf, it is as if someone has taken this sense away – you can now only see 180 degrees on the horizontal and 100 degrees in the vertical – what we call your field of view. However, studies have shown that those who cannot hear are much more sensitive to moving stimuli in the periphery of their vision; their visual attention is heightened. Without getting too deep into the neurology and the current research that is helping us understand how our eyes and brains work together, let's look at it from a rewilding point of view.

I'm going to assume that your eyes are healthy, that they work, and that if you need corrective prescription lenses, either glasses or contact lenses, then these are up to date. Otherwise, no amount of awareness trying will be able to help you out.

When we look at a scene, our eyes are flicking about all over the place, scanning what is in front of us because we have a fovea, a relatively small area of the retina on which to pick up high-resolution information. It means that we have to flick it around a bit, rotate our eye in its socket by tiny increments, training this highly sensitive part of the retina on various points of interest within our field of view. Thanks to the power of our mighty oculomotor system, we can make these discrete shifts of gaze two to three times every second, something known as a saccade. The eye then pauses on the points of fixation before jumping on again. This is part of our exquisite design and it's what we naturally do. So much so that staring at a single point and fixing on it for more than a few seconds feels highly uncomfortable and exhausting.

Lift your head up . . .

So how do we become better at looking? Well, the first thing is to actually hold your head up. I tend to watch people as I would an animal; it's something I do, I can't help myself, it's a learned behavioural trait that delivers for me in my work and pleasure.

The first thing a naturalist notices is how an animal or bird, or even a tree, holds itself. This gives us a massive clue to its identification.

It's the 'J' in the phonetically and etymologically confused word 'jizz' – which is something a bird, butterfly or dragonfly watcher might use to instantaneously identify a species of their quarry when they clap eyes on it. One theory as to its origin as a word is that it originated as an American acronym for identifying enemy aircraft and stands for General Impression, Size and Shape (GISS).

What I'm getting at is that I've been noticing something, something that is slowly becoming more entrenched in our own daily and social behaviour, an insidious creep of technology into our daily norm. Ever since phones became mobile and no longer connected to a wall with a curly wire, they've stolen our sight. We're going blind and the phone is our technological glaucoma. Our eyes are our primary sense; we've evolved to be stimulated that way, but they can't be looking everywhere at the same time. We're no longer 'saccading' over a bigger scene; we're being mesmerised by the LCD screen.

As a species we are suffering from a collective tunnel vision. At a bus stop in Lhasa, Tibet, every one of the eighteen people I stood with was gazing, mesmerised, at a retina phone screen. Downtown, even the older generation, dressed in traditional sheepskin *chubas*, could be seen in doorways or leaning against temple walls, their faces illuminated by the cold blue glare of cell phone and tablet – even here, where nature culture is still close to the surface, phone culture is prevalent.

Several weeks later, I took a short walk into the city centre of Winchester – a quick nip to the shops to pick up a pint of milk and a pasty for my lunch. I thought I would do a quick survey of all the people I passed on the street. Of the eighty-seven folks on the pavement, forty-six of them were looking at a screen.

It's no wonder that when I talk in the pub about the pair of peregrine falcons that I see regularly roosting on the cathedral and the urban drama when one of them takes out one of the pigeons in a puff of feathers right in front of the busy comings and goings of city folk, people look at me in disbelief. Most aren't aware that this bird shares their city. It's why these observations of naturalists can seem so other-worldly and fantastical, as if we're some kind of guru trained in an exotic, mystic skill. We're not; we just have our eyes pointing in the right direction most of the time. Scary proof of this modern myopia is the town of Bodegraven in the Netherlands – here, at pedestrian crossings at traffic junctions, the local council has installed light strips in the pavement that change colour. Red for 'don't walk', green for 'go' – all so that those itinerant mobile users don't need to lift their eyes from their screens in order to see whether it is safe to cross or not. Is this the future?

I partly blame this handheld technology, although our perceptions of the natural world are changing at other levels too. The innate need in all of us to be needed or loved means that we scan our screens almost constantly, filling up 'looking' or, more importantly, 'noticing time' with visual information that isn't directly relevant. With all our attention focussed on what is immediately in front of us, the delicious, visually sumptuous backlit graphics, we are rolling down the blinds on our window onto the real world. It's all i-Phoney; we're staring down a never-ending digital tunnel, never quite reaching the dazzling promise at the end.

This is not just a behavioural change in the city either, I see it in the countryside – people walking their dogs, but not watching their dogs or, indeed, enjoying the scene they are romping through. This technology-driven crumpling of our posture means that our heads are down. Runners may also suffer from this head-down position as they look down at the ground immediately in front of them – nervous about what they might step in perhaps? This has all sorts of knock-on postural issues which create tensions in the neck and back. But for now we'll just tackle its influence on what we see.

Now compare this modern domesticated (in the living-in-a-built-up-environment sense of the word) world with those of native cultures, people still living an eco-centric lifestyle. They look up, they scan the distance, looking for a prediction, a clue as to how the future is going to shape up, from incoming weather systems, which have an immediate implication for all manner of activities, they look for clues about food, plants and animals that can be eaten and, more importantly, they check their social status, and give each other updates, assessing how they are received by friends, family and associates by looking them in the face. Compare this with the married couple sitting at a restaurant table gazing lovingly into . . . their phones.

You could argue that the need to look out for predators, or for dinner, has been functionally replaced by the equally significant modern need to see whether your boss has called, whether you got the job, or if your bid on eBay bagged you that bargain. However, our bodies don't know this; our intellect has, once again, outsmarted our physical nature, our wild needs.

In our natural form, we are not designed to spend hours and hours, days upon days, sitting in front of and scrutinising near objects. We are supposed to be looking up, looking out – looking out for a threat or for food. It's a much more natural posture.

Slow down – have yourself a look, see and listen

The next thing is to slow down. There are many reasons for this, but the primary one is that if you are moving more slowly, you simply give yourself a fighting chance of noticing the details of what's around you. It sounds so obvious but in an increasingly abstruse world looking for complex answers, sometimes it's the really simple stuff that is so easy to overlook.

My first ever trip to the Neotropics was a case in point. This wasn't just my first trip to South America; it was my first ever trip to any non-European country, the first time I was to experience the bountiful biodiversity of a tropical ecosystem. As I grew up during the seventies and eighties, reasons for 'saving the Rainforest' were well known to me. It was the 'lungs of the world', home to 1,300 of the world's birds, 430-odd species of mammals, at least 450 reptiles, an estimated 2.5 million species of insect – and the musician Sting. I had been given the opportunity to experience the exaltation of life in all its plenitude. I was more than excited.

The flight got into Georgetown, Guyana late; it was as dark as the hides of the cows that lay on the road, the same dozy bovines that the taxi driver neatly swerved to avoid on at least three occasions on the short trip to the hotel. En route, the sounds of a tropical symphony hung in the heavy air. For a young naive naturalist, the stridulations, creaks, whistles and farts of countless crickets, cicadas, frogs and nighthawks leaked tantalisingly in through the cab's windows. I was bursting with anticipation, it was a real dream come true, something I never expected to experience, even in my wildest biophilic schoolboy fantasies.

The following day, before dawn, I was on the flat roof atop the main big city hotel waiting for my first eyeball full of tropicana. It started well: dozens of birds of every colour started waking up and decorating the dawn. They were mostly tanagers, but a rainbow of feathers, and I was reduced to shouting primary colours to the hazy rising sun. However, my joy was short-lived.

As soon as I headed out away from the capital and into the wilds and found myself immersed in the primary forest of the Guyanese Rainforest, a green curtain fell over my enthusiasm. There was, it seemed, nothing but a palette of every kind of green and brown known to the universe, as well as mosquitoes.

Where were the snakes coiled around every tree? Where were the multitude of vibrant parrots, the dozens of butterflies, the frogs, the lizards? Where was the party? It felt that I had been fed a lie all my life. I remember settling myself back against a substantial buttress, feeling more than a little disappointed and underwhelmed.

But the forest felt my disillusion and, first, it presented an ant. A tiny offering of tropical biodiversity, but a morsel of animation to catch my eye, tether my conscious thoughts and focus my attentions. Fortunately, this ant was not one of the more plain or difficult species to identify, with which the tropics are blessed in plenitude, but a leaf-cutting *atta cephalotes.*

This little golden girl was making small purposeful strides down the razor-backed horizon of the buttress, carrying her heavy load, a disc of leaf, measured and cut to optimum size by the sharp calipers of her own jaws.

This ant species was a well-known character to me, a super-star of the myrmecophilous world, a pin-up for the myrmecologist and a species of fascination in my bookish childhood. To meet this insect in person was bringing the pictures on the pages of my encyclopaedias to life. She tottered off across the forest floor, green flag held high. As I followed her with my eye others seemed to materialise – an identical, leggy sisterhood all heading along the same path. But they had been there all along. It was as if my calm, respectful, patient presence was being rewarded.

The longer I sat still, the more I was gifted. Other ants of different designs appeared: soldier ants, thick-headed brutes armed with scimitar jaws and guarding not one but several scintillating tentacles of ant highways, now seemed to disclose themselves while others headed in the opposite direction to the first. The forest was alive with them; every turn of my head was met with ants, ants and more ants. Then a procession of things seemed to come trooping out from the wings – the green curtain was being lifted slowly and, in the next hour, the theatre of the forest was revealed. Hummingbirds dazzled in the spotlights of the sun's rays filtering through the canopy; a parrot snake dropped its jungle vine facade; a burnished feathered jacana armed with a rapier bill cut a dash to skewer butterflies. After I had absorbed as much as my senses could process and still retain some meaning, I got up and wandered back to camp. As I did so, it was as if the world had sped up – as I purposefully stomped off down the path, I slowly become less and less aware of any animal life. An inversely proportional relationship between speed and the creatures seen. Progress and experience at odds with one another.

If you remain stationary and quiet for long enough, after a few moments, those eyes present when you arrived will, with no threat perceived, carry on their business. Others new to the scene will appear, and you will become no more of a threat then a wormy dead tree stump or fence post, accepted almost as part of the scenery. If you sit still, like a pop-up book, life will make itself known as you gradually tune in to your environment.

This is an illustration of a technique in ecological immersion that I've been practising for years, at first without even knowing it. Initially, with a fishing rod in my hand, it was a pleasant side-effect of angling, something to distract me in between bites, but over time, it became a necessity, an activity all on its own. The faster our lives become, the further from our start we get and with that the ability to be, the purpose of being, stationary becomes more and more distant. So while staying put and not moving is a sort of default setting for our nature and therefore our relationship with nature, it becomes a switch we rarely get to flick. It's seen as a waste of time, doing nothing when, in reality, it is an opportunity to be doing so very much more. It is noticing, it is beginning to see and, when you can see, you start to connect.

Open your eyes

The ability to read a landscape, to carefully and consciously look over everything, to search as much for natural patterns as for things that don't fit those patterns, to know how to scan the landscape for signs of the enemy, is considered an essential and basic part of military training. It's not, if you think about it, anything new. Again, it's a basic skill-set that our ancestors and those still living directly from nature utilise to this day. It is a skill that you can work on and develop yourself, but it requires focus.

As you read this, you're reading from left to right; it's a habit you've been taught and you've almost certainly got into the habit of skipping words and predicting the placings of verbs, adjectives and nouns, understanding the sense without necessarily tracing every letter or syllable. Your eyes are jumping ahead of what your thoughts are processing. However, try reading backwards. Start on the right and scan along the line to the left – you slow down, you're swimming against habit and have to pay much more attention to each letter and sound; you have to really concentrate.

This is an amazingly effective little trick with which to look at the world. Once you've retrained your brain to be able to do this, to be super-aware, to really look and see every little detail within the limits of your vision, you can start operating in all directions. Combine this with the 'sitting' still technique. Find yourself a spot outdoors, as wild as possible, and settle yourself down comfortably. Now work the landscape in front of you from right to left, run your eyes up and down every line, imagine you are sketching the scene with a pencil and let your eyes trace these same lines. Look for things, give yourself tests – how many living creatures can you see, imagine what the trees, stems and leaves feel like, mentally crawl through the scene, turning stones.

Expanding your view is not something you can physically do anything about; your field of view remains fixed, that's the physics of the situation. But what you can do is try and train yourself to be more aware of what it is and its limits. Explore your visual periphery. Work on your visual attention. Hold your arms out straight at shoulder height and wriggle your fingers while looking straight ahead. Adjust with forward or backward motion the position of your arms until you are aware of, but not looking directly at the movement of your fingers on either side: that is the horizontal limit of your field of vision. Now do the same in other directions, up and down, and every other position in between. You've now felt with your fingers around the very edge of your visual field. It's a very physical demonstration you can do with yourself that gives you a clear idea of how big a space you can scan.

It's similar to an experiential game I've played with kids to help them get better at being visually aware of their wide-angle vision. I call it 'Being a Buzzard', though of course you can substitute any bird of prey. The basic idea is that your outspread arms are your wings and you have to fly around until you find a perch (seat) overlooking something you might find interesting, then you sit there really, intently exploring your domain. You look around it, imagining that you're a bird, looking for prey, thinking about the habitats and what animals might live in them, using the awareness of your full periphery to check and be aware of the trees, clouds, grass, hedges and herbs moving around in the breeze. Once the children have played around with this a bit and expanded into their full visual panorama, I tell them they can start hunting, trying to find the movement of birds, insects, anything that might be prey to a buzzard.

While carrying out these exercises which will start as being real, consciously focussed efforts, you'll almost certainly fall into the trap of concentrating too hard. Remember the field of vision's natural dynamic state – your eyes need to dance about a bit; if you try and fix on an object, you'll get tired and part of the idea of being aware is being at peace and relaxed with yourself before you can be expected to become comfortable with your surroundings.

I hinted at this earlier on. The processing of the image and the ability to superimpose and use your imagination based on experience is an accumulative thing. When a child has his or her first experience of something in nature, it is full of a primal fascination, wonder and curiosity; it's the same experience repeated by generations of hominids, it's the exploration of the environment at its most basic.

When I look at a scene, I like to imagine that I can zoom in on things that I can see. If I'm looking over a steep-sided valley at a hanging western oak woodland, sparkling silver in the spring, in my imagination I become Ant-man: I zoom in, I let my imagination carry me in an instant to the other side of the valley, and when I'm there, I land and climb over the textured surface of a lichen as if it's an undulating hillside. I imagine the herds of microscopic insects, bark lice like micro-bison on the prairie. The reason this is possible is that on some past adventure I have explored that other side. I know what grows there; the redstart that nests in a knot hole on an old planted sycamore; a dead oak limb which sticks up at a forty-five-degree angle and which, last year, harboured a nest of lesser-spotted woodpeckers; the dormice in the thickets of the hedge; that there will be caterpillars of pearl-bordered fritillary butterflies nibbling the violets' leaves while they peak their purple heads through... the list goes on. It's a world of interconnectedness and again, simple though it is, every single experience, no matter how brief and fleeting or incomplete it may feel, has a place, adds a pixel of detail to the understanding of your patch, your environment.

6

The Blind Birdwatcher

'THERE'S A bullfinch coming, it's about to pop out of that gate, a sparrowhawk's in pursuit, five, four, three, two, one . . . there!' said Gary. And, sure enough, a male bullfinch came bowling through the gap in the hedge, a male sparrowhawk close behind, full stretch, keening at the finch's tail feathers just as Gary had predicted. My eyes (or, that should read, my ears) were truly opened to the untapped potential of our senses when I went out birdwatching with a blind man. This oxymoron of an experience happened when I had been out recording an item for a BBC radio show. I have recorded hundreds of such pieces over the years but this one had a lasting effect on me, one which I use every single day.

We had arranged to meet up with Gary at a prearranged location, a field centre, buried in the ancient folds of the Dorset countryside. It was a verdant and complex patchwork of greens of at least a dozen shades separated from each other by frothy hedgerows, fizzing with the pink, delicate, aromatic blooms of quickthorn and the lime-tinted pads of elder, which, from a distance, looked like trails of foam.

This terrestrial feast for the eyes was further enhanced and complemented by the azure sky and its flocks of sliding white clouds. Birds were in full procreative swing and the bees, having taught them all they know, were at it themselves, throwing themselves, with the rest of the pollinating insect hordes, at any one of the many hedgerow blooms.

Initially, in my gauche naivety, having met Gary in the car park of the field centre, I experienced a pang of sadness. Here I was with a fellow human, a biophile like me, a naturalist who wasn't able to witness the English springtime in all of its fecund frenetic fornication. For Gary there was no electromagnetic stimulus, no potential difference firing the neurons in his optic nerve. Which, quite simply, meant no flowers, no diaphanous damselfly, no lazy green-veined white butterfly, no inflated bullfinches bursting with territorial pride, no nothing, as far as I could see.

But, as it happens, I couldn't see very far at all. Like most of my species with a fully functional quota of senses, I was simply not using them right. I was about to be shown up as being decidedly short-sighted. The bullfinch and the sparrowhawk were not only identified by the sounds they made. A combination of an accurate mental map of his surroundings and an instantaneous mental overlay of other more subtle audio clues were brought together to give a very clear notion of what was going on.

Gary explained to me that the alarm call of the bullfinch was the starting point, then the alarm calls of the swallows and house martins that cut sweeping dashes high above our heads told Gary that there was an aerial predator around (like many birds, they have predator-species-specific alarm calls); the short, staccato *flitt, flitt, flitt* of the barn swallows was enough to identify a bird of prey. This and the Doppler effect of both the bullfinch call and the audio wave of the hirundines overhead tracking with the predator, plus a bit of knowledge as to how different birds hunt, gave Gary the rest of the picture. This was birding beyond binoculars. Amazing though this demonstration in refined listening was, I also had to acknowledge that his was not some superpower. Gary wasn't wearing any kind of Lycra and yet his perceptiveness of the world around him was to me akin to *Marvel's Daredevil.* It got me really thinking about what it was I was regularly missing and from that moment on, I've endeavoured to improve my sensory awareness of the world around me, wherever I am.

Up until that point I was like the majority of us, bumbling around in the world, my eyes doing most of the sensory donkey work. As I am a visually oriented creature, it's easy for my other senses to be swamped, their input not to be logged, the neurons they fire off to be ignored, their sensations never registered, in favour of our overly obsessive orbs. It was as if my eyes had led my other senses astray. I was seeing at the expense of hearing.

It's easy to believe the overwhelming, overruling view that our sense of hearing is rubbish. The textbooks tell us lots about bats and cats, dolphins and dogs and their incredible audio responses to sound. It's true that all of these animals have unique adaptations and abilities to perceive sounds that are out of our audio range, particularly in the higher-frequency register; a fact that we'll come back to a bit later on, but one that is useful to keep in mind at all times when you are being in nature.

Most of our modern lives are full of noise, our ears are stuffed with and over-stimulated by sounds of our own generating. As I'm trying to write this, I'm being audibly distracted; my ears are assaulted by sounds I don't want to be hearing. They're being bombarded by tree surgeons working on a neighbour's overgrown hedge, my wife kicking the hoover into action downstairs, the computer fan making a whirr, the hard drive ticking at every input of the keypad, and although there are birds outside my window, continuously communicating (I can see their chests puffing and their bills opening), I can't hear them very easily, not unless I make a conscious effort to focus in on them.

A modern human accepts the background audio wallpaper of their everyday existence almost without question. We've acclimatised ourselves to accept the sea of white noise that is a product of our contemporary lifestyle, our cocooned and sheltered world of our own making; the shuffling of flat feet, the rustle of synthetic fabrics, jingling zippers and electronic notifications. Noise cloys to humanity. There are many tiers to this audio obstruction between us and nature.

With around 50 per cent of humanity living in an urban civilised environment, that's half of the population that is less connected to the natural world and a natural experience of it, obstructed and influenced by anthropogenic factors. How we relate to the world via sound waves is underappreciated. Most of us are not aware of our personal audio pollution, a fug of sound that follows us. As a country boy, I find simply sleeping in a room with the incessant background thrum of industrial air conditioning units impossible, it invades my primary audial cortex, and even this mechanical murmur is drowned out by a yet higher tier of decibel-producing machines, the permeating rumble of rubber on tarmacadam or steel against steel.

So, is it any wonder that when first we're exposed to a natural soundscape, we're so deafened by the silence? We don't perceive how noisy we actually are or the sonic signature of everything. The wind moving through the different species of tree, needle, leaf, bare branch or full foliage, or the more subtle soundings of a bird's syrinx (more on what this is later), are sounds lost to our modernised ears.

We need to become more aware of our own personal acoustics, not only to aid our observations but also to gain valuable information about our immediate environment.

It's a really difficult message to get across. When I accompany others out into a field situation, rarely am I pleasantly surprised by their field skills – whether it's trying to get into intimate proximity to timid and highly strung animals like pine martens or deer, leading a dawn-chorus walk, or simply going for a nature walk. The first thing to tackle is the personal noise.

Choice of clothing is part of this. Obviously, if we all walked around without a stitch on, we would be as nature intended. I suggest you try it sometime, it really does make you appreciate what an audio ball and chain the draperies and decoration of everyday life can be. True, this is probably taking your personal rewilding to a level you might not be prepared to go to, and, indeed, it is rarely socially acceptable – unless you are a naturist naturalist (a sentence guaranteed to confuse). However, there are ways and means to explore this which are outside the immediate scope of the book. But you get the point. Our wrappings hinder us in as many ways as they help us.

The best compromise you can make is to wear non-rustling clothing; there are plenty of options, from clothing made from natural fibres to those with a waterproof barrier layer integrated in such a way that it isn't on the outside. Those who spend a lot of time out of doors often speak of layers, with the shell layer being the outermost one. Sure, this has a function, but it is often this waterproof and windproof shell that creates the biggest din. Try walking with someone dressed from head to toe in Goretex. The swishing noise made by thigh rubbing against thigh or the enthusiastic swinging of the arms is at best mildly irritating with its high-frequency range and rhythm, interfering with the incoming cues coming towards your ears from the environment. But, at worst, that same 'whispering, walking sound' is being emanated in the higher registers. The same frequency that many other creatures reserve as their emergency bandwidth and to which they are highly tuned as a necessity of survival. Many alarm calls of birds and mammals have high-frequency components to them. Have you ever opened a packet of crisps or unravelled a sweet wrapper in the presence of a sleeping hamster or rabbit? If it doesn't immediately awake, it certainly flinches. These high-frequency alarm sounds are designed to not travel too far – that way they can have a directional component to them. The sound says here I am, this is where the threat lies.

When I sit quietly in a wood, I can track with my ears the progress of a predator, often a dog that has slipped its lead, simply by the high-pitched alarm calls of robins, wrens and blue tits. A well-known birdwatcher's trick to inveigle a skulking warbler to show its face for long enough to be identified is a noise that goes something like *pisshhtt* – in essence it's an alarm call that makes the bird pop up, out of preprogrammed necessity, an instinctive need to know, before it scarpers into the thicket.

This is pretty much the effect that many of us obliviously have on the natural world around us; we are a walking, and often talking, alarm call.

So the first thing is to dampen and eliminate that rustle. It's the same rules as in the previous chapter on how best to notice things, and it goes back to the speed with which you are moving: the more haste you make, the more noise you emit; it's as simple as that. Theoretically, it's possible to be wearing a bin liner and move like a fairy's breath if you do it slowly enough.

Simplify your profile. Once I met a couple who bred llamas. Their animals were their family and they would come into their cottage kitchen, a warm and welcoming, cluttered room, with dressers piled with plates and other crockery, bottles and jars on every surface, and yet the llamas would come in and walk around, even with packs on, and not knock a thing off a single surface; they seemed to have an uncanny highly developed sense of personal spatial awareness. Be a llama. You can help yourself by simplifying your profile. Carry as little as you can, avoid too many things with straps that can snag. If I can, I try and get everything in my pockets and therefore choose field jackets which are well provided with storage solutions. Backpacks are best if they are no bigger than the size of your back and as slim in profile as you can get away with.

Be aware of the noise that fastening might be making as well. Zip pulls are the most frequent offender, a small clinking bell, announcing your presence to all and sundry with sensitive ears. Poppers are best kept closed or they might knock together, and if you've got Velcro – well, just carefully choose your moment when you decide to pull the two fabrics apart.

How you place your feet without sound is certainly a habitual movement you need to develop. I've been fortunate in that I've spent quite a bit of time with various hunting-gathering cultures around the world and something they've all got in common is how they move through the landscape.

This first came to my attention several years ago, when working with the Maasai people. I was filming their extraordinary symbiosis with a small bird called a honey-guide. The bird alerts the villagers to the fact that it's found a bees' nest and then it leads them through the bush with a distinctive 'follow me'-type contact call. So we followed them, following the bird, with the hope of discovering a wild bees' nest – at which point, we'd have the punchline to the film narrative, the bird would have its bee larvae and all the wax it could consume and the Maasai would have their sweet gold.

All was going well, as we snaked our way through the crispy, crunchy dry bush of this part of Kenya, until a small hiccup in the communications between the bird and the Maasai occurred. The bird had, in its over-eagerness to get to the nest, dropped us, or it may have been the fact that a camera crew with all its incongruous kit, weight and bush-tangling cables had slowed the expedition down. Whatever the reason, the bird had either got ahead of us or changed its mind and gone quiet. Our Maasai friends, seeing how hot and slow we were becoming, told us to rest up at the side of the trail while they went off to find the whereabouts of the bird.

Twenty minutes later they found me sitting on a dead tree, head in my hands, quietly sweating cobs in the shade as I awaited their return. Without hearing the sibilance of a single dry leaf, or even feeling the displacement of the dry air, there was a tap on my shoulder. They had relocated the bird, but the honeyguide and its special relationship with the Maasai wasn't the only thing now on my mind. Something else, much more useful to me, had piqued my interest. I wanted to know how our guide had managed to move so silently? How had he executed his catlike approach? I made it my business to find out, I wanted to own the secret to his spectral walk.

The ground was carpeted in a biscuit-litter of leaves, each one as dry as old bone, without a molecule of moisture to quieten their complaints when trodden on. It was like trying to walk silently on potato crisps, without making a crunch and without losing speed.

My first question was why? Why did he feel the need to walk so quietly? It wasn't as if the bees would hear him coming and buzz off, and the honeyguide was positively craving his presence. The Maasai would also occasionally quietly whistle and talk to the bird, a way of reassuring it we were still coming, and that we were still intent on upholding our part of the deal.

My Maasai guide told me that it was simply what they did, a good habit to get into; then, after a little more thought, he added that in a land where there are large predators and other creatures that could cause harm to your person, being part of the background, a no-noise in the natural soundscape, was how you avoided trouble. This ability to not attract unwanted attention to yourself is a fundamental part of knowing what's around, and being aware of what else is walking the bush and treading the soil with you. It is a vital part of everyday survival. It's not simply about hunting, it's an important part of not being hunted. The aim is to be at one with nature, to sound like the wind, the rain, the trees, the animals, the birds and the insects. This way you don't attract attention and the danger that can sometimes come with it.

The method my guide had was very much the same as that of almost any other indigenous person I have ever spent time with. Various Native American tribes, the Jahai of Malaysia, the San and Himba people of Namibia, the Irula in India and the Makushi of Guyana all walk with the same careful consideration. Barefoot is best, but many were nearly as effective in their sandals or flip-flops – though several would remove these for the purpose of stalking or walking towards quarry only to reposition them on their feet when they had finished their business and they were relaxed and off duty. Here in the Western world we tend to do exactly the opposite.

A (very) quiet walk in the countryside

The basic technique is again centred on full awareness and focus. You think about walking and you tend to think of feet and maybe legs, but a good wild walk is done with the complete coordination of your entire body; it's as much about balance as it is about where and how you place your feet. So with this in mind, let's start about as far away from your feet as you can get.

Breathing steadily, through your nose and from your diaphragm, is as much part of this as are your feet. A technique familiarly practised in the martial arts, its purpose is the same. It keeps you calm, mindful and at peace and helps you focus and move with famed stealth. It also has many advantages over the other option of breathing through your mouth: you can engage your sense of smell, your throat is less likely to become dry and tickly, you conserve more moisture (important if it's hot), and your nose naturally filters debris and dust which might cause you to cough (and therefore make noise) if it was to lodge in your throat. In addition to this, a sneeze reflex is generated from the nose and can, in most cases, be suppressed by pressing down on your top lip with a finger (this intercepts the nerve pathway that set up the sneeze reflex in the first place, a sort of neural short cut) or at least muffled, whereas a cough, on the other hand, usually originates from the throat and is something that you can't really control.

Your feet should be the main points of contact with the environment and therefore how to place these on the ground is vital to successful quiet walking. As you move, you have to be aware not just of obstacles, but of the substrate you walk on too. Choose the path least likely to crunch or rustle. If you're walking along a woodland path, step on the green stuff; grass, moss or clear soil is preferable and mutes your steps while step-amplifying dead vegetation and loose gravel and stone is to be avoided. If you're off the path or have to move through vegetation, choose the clearest route through, picking up and moving any obstacles likely to snag or catch.

How you step is obviously quite important. Your tread should be slow and deliberate; the full weight of your body is not brought down in one go, but slowly applied in a roll, with no lateral foot movement at all – any slip or slide creates noise. Sometimes it's helpful to think of the economics of moving while you walk. Noise is created by a superfluous movement, its energy wasted as sound energy. So if you are moving with your ultimate efficiency, you shouldn't make much sound at all. A good practice is to try walking slowly with your shoe laces untied – any dragging of the lace will emphasise any sideways movement.

Imagine peeling a flat foot off the ground by the toes, but in reverse. Heel first, the foot is rolled into place, through the ball of your foot to your toes last. If you're stepping backwards, it's the same technique in reverse, the toes first, then the ball of the foot and heel. This rolling, peeling walk can help you move quietly over the most potentially noisy ground. If you place your foot carefully and gradually, you can feel if there is anything under it that might sound, such as a twig or a particularly scrunchy leaf, in which case you can abort the step and relocate it to another spot. It's key to not put all of your full weight down on your leading foot until you are sure it's a good step; in this way you avoid a heavy thump and you are more likely to feel any noise-prone obstacle underneath.

It can help a lot if you bend the knees and, when you're getting very close, you need to slow right down, making tiny steps. Your footsteps should fall just behind each other. You can also minimise the risk of noise even further and go into super-stealth mode, which entails the same rolling action but this time using the outstep of your foot, effectively walking on the edges; it's quite uncomfortable for long periods of time, but for a short final creep in close, it's a handy technique to know about.

Another trick familiar to those who watch cowboy movies is the cross-walk, this is handy for moving a little faster and involves standing perpendicular to the direction in which you want to travel. With feet slightly spread and knees bent, swing your following foot in front of your stationary one, then, when this is paced, swing your following foot back around to the starting position.

I have come across variations on all of these themes. The 'fox walk' involves not putting any weight on the front foot until it is completely placed the ground. The outer edge of the foot is placed in contact with the ground along its whole length. Then, when it's in position, the foot rolls sideways so that the rest of the underside of the foot makes contact with the ground. If all this is done, before the weight is placed on it, then it is possible to reposition the foot in a better place. If you keep your stride length small, you'll have quite an arc of space available to reposition your foot within.

The 'weasel walk' for close approach and stalking involves curling your toes up and placing the outside ball of the foot down first, then rolling it down towards your instep, then the heel, then the toes. The 'cat walk' involves pointing your toes towards the ground on almost cartoon tip-toes and then rolling from the edge inwards, before bringing down the heel; this is done really slowly, while all the time feeling for obstacles and things that might produce a sound and give away your presence.

* * *

Try to be aware of the rhythm of your movement; there is no sound that stands out from the random tangled chaos of sound, the wind in the trees, the rustle of leaves and the gurgle of running water, so much as a regular beat of footfalls. So try and break up your natural pace with regular stops and pauses as well as changes in pace (also useful for getting quick bearings and sounding out your vicinity).

There is no greater demonstration of the connectedness of human nature than that seen when out with a monkey-searching party with the Penan people in Sarawak or when tagging along on a woodland-management deer stalk in Hampshire, though the two are separated by over seven thousand miles.

The environment and the quarry as well as a substantial difference in the amount of clothing each hunting party is wearing is very different, but what they share in common is the way they move and collaborate and synchronise their movement, in order to achieve the same goal: namely, to outwit a creature of prey, and one that has an ear open for the movements of a predator. The rules of the game are the same.

When in a group, another factor needs to be taken into account: you need to be in unison, and the dance becomes one that is choreographed by the leader. You have to become much more aware of all the others in your party. This means primarily matching the footfalls of those in front of you. By timing the moment your foot is placed on the ground with that of the person in front of you, effectively you are creating just one audio disturbance at a time, one set of footfalls to be heard, if they're to be heard at all.

When walking in sync like this, try and notice how the ground responds to the person in front. Does it sink? Squelch? Did it make a noise? If it didn't, try literally stepping in their footprints; they've tried and tested the ground for you. If it did, then try and find a better option, but all the while try to match left with left and right with right, shadow the cadence of your leader.

Recently, how automatic and first nature this way of walking becomes was illustrated when I took an enthusiastic group of young college lads from Winchester out into the New Forest with the stalking team that manages this particular woodland. The aim was to try and get them to ringside seats at a rutting stand of fallow deer, to utilise these basic stalking techniques in order to get close to these large and impressive mammals as they are engaged in one of nature's most spectacular breeding displays. However, when we set off, I realised that we had to go back to basics. I hadn't appreciated just how inefficient we can be with our footfalls. Every time we stopped or paused in the woods, trying to listen out for the guttural grunting of a rutting buck, the boys kept moving. While we were no longer making forward progress, the boys didn't stop moving their feet; they took it as a time when nothing was happening, so they kept on kicking about the leaves. Each was foot-fidgeting to his own rhythm and the group was unwittingly creating a sound blot on the woodland soundscape, a din of shuffling dissonance – all without really being aware that they were doing it. After a few basic lessons, they soon got the hang of it and with real silky stealth and economic footfalls, they took off on their own – from undisciplined naturalists to making contact with the innate hunter-gatherer in a matter of minutes. The result was a close encounter and unforgettable moment with a spectacular mammal at its finest, a life-long skill and a deep-routed connection with part of themselves that had been dormant up until this moment in their lives.

One final thing: while all this seems a lot to concentrate on, you'll start off consciously going through each step one at a time, which is in itself a healthy, mindful thing to be doing, but in time this will become second nature and, if used a lot, first nature; it is, after all, how we are meant to be – all you've done is make contact with a long-lost you.

Play around with the various techniques to find one that works best for you in a variety of situations and, in addition, be aware of the fact that you are never the only thing out there making a noise, which is the whole point of this chapter. Listen for patterns in the ambient soundtrack of your chosen habitat. Move when the wind blows, or a plane flies over, and you'll mask your own sounds and thus use the acoustic cover they provide to your own advantage.

7

Eavesdropping on Nature

WHEN I was a plastic-minded impressionable boy, I had a vinyl record. I wish I still had it so that I could tell you what it was called. But it was a recorded story, voiced by TV presenter Johnny Morris. I loved Johnny – he presented a kids' animal show, *Animal Magic*, and this record similarly played to this persona. He was telling a story in the first person, of him waking up in bed and listening to all the sounds he could hear while lying there before getting up to start the busy day. What made this such enthralling listening for me was that this imaginary street was populated with animal families and he then proceeded to describe what they were all doing via a collection of recordings from what was probably the BBC wildlife sound archive.

The sea lion family next door, with Mr Sea Lion gargling in the bathroom, an anthropomorphic image conjured up by a real-life-recording, and the fallow deer across the street, trying to kick-start his scooter; to this day, I just have to hear the burbling, coughing, throaty roar of a fallow buck, calling in the rutting season, and I get a mental image of a deer sitting astride a 50 cc Vespa, desperately trying to kick it into life.

I would often find myself lying in bed doing pretty much the same thing, just without so many exotic animals to listen out for; there was something comforting – and there still is – about getting a handle on the world via your ears before you get up and stagger into it. I recall lying under my duvet, the pale-blue light of dawn just getting in around the curtains. I would listen out for the milkman and the purr and clink of the electric milk float. I would listen to the milkman's footsteps and try and count the number of bottles being picked out of the crates, and then in my mind's eye try and link the count to the bottles I would see on doorsteps on my way to school. I'd listen to my dad get up and go to work, the metallic fumbling, of, first, the keys, then the door handle, then the starter motor – four turns before the Vauxhall woke up – and then I'd hear the purr of the Viva's 1.8-litre up the road, Dad working his way through the gear box. This was the dawn chorus for the earlier years of my life, inspired by Johnny.

Then when we moved to the countryside proper, my ears popped, as if they had been full of water, and suddenly cleared. The music of nature was much closer to my window, which, together with the old single-glazing, no curtains and iron frames, meant I could really let my imagination fly out of the house to the sounds I heard. Here the dawn chorus was less about machines and the trappings of an urbanised world of tarmac, concrete and machine. It was as it should be. I really heard birds for the first time. As before, I would lie in bed but I would layer the familiar household pre-dawn sounds with those characters sitting in the honeysuckle, oak and thorn.

The mighty wren, an angry little man, the ticking alarm of a robin, the freewheeling bicycle, the dark cowl of the blackbird, his song as rich and deep and classy as his plumage. I was beginning a process, building a picture, telling myself a story, getting to know the characters by their acoustic signatures alone. Unconsciously, I had started practising the beginning of what some would fancifully call 'mindful listening'.

In the same way as a person can recognise the voices of those people familiar to them, the voices of their parents, brothers or sisters, I could recognise the song thrush, repeating his phrases, as if trying them out to see if they fitted the melody, then getting bored and discarding them for the moment. I could hear the cock bird at the top of the big oak and I could also recognise his neighbours, rivals for the attentions of the hen birds. In other oaks across the field, and still further away, there was another of the same kind singing in woodland by the old sandpit by the badger setts.

The thing is I didn't know what they were at this stage, I hadn't watched the spotted breast swell with its music, and similarly, all the other birds of this acoustic landscape were still strangers as far as spotting them was concerned. But that didn't stop my winding thoughts and imagination tangling with their melodies, I was really listening to them, trying to make out each and every breath that vibrated through their windpipes. I was telling their stories as I listened to their comings and goings, picking out the differences between their various different vocalisations, trying to match the song, the contact call, the alarm call. I recognised individuals, too; that song thrush across the field had a few unique inflections and a couple of phrases that my neighbouring one never used. I imagined how and where they were perched, their size and what they might be communicating to each other in the twilight. My imagination was run through with the song of their lives. I had built the beginnings of an acoustic ecology, started to notice the players without so much as setting my eyes on them. It was rich pickings for my hungry ears, and it still is.

The practice of listening deeply to a landscape, not just birds but the whole interconnected ecology of sound, is such a vital part of our natural connection. The wind playing across the hills, tuned by the species of vegetation it encounters on its way, each grass stem, twig and branch, a reed sounding and lending its own individual qualities to the acoustic landscape. Add to this the animated minstrels: the stridulations of grasshoppers, the scrape of lizard or squirrel claws on rock and bark, the mating music of birds, amphibians.

There are even sounds quieter than the whine of a gnat's wing, secret sounds easily missed: worms, pistons in their burrows, gurgle with the push and pull; barnacles fizz and pop; snails rasp; place your ear to the hulk of a dead tree and you can almost hear the decay; the scouring of countless beetle larvae jaws; listen attentively to a pond and you may hear the common but rarely heard singing of water boatman beneath the surface.

Each of these parts adds richness to the world. They are instruments in the whole, the players in the orchestra that gives everywhere a sense of sound. A unique web of noise, a sound sense of place.

Just as many self-help gurus preach the art of listening to strengthen your business relationships and increase the effectiveness of your dialogue, we can apply the same to our relationship to nature. We're all in a dialogue with nature, a conversation of some kind, but it may not be one in which we're doing much of the listening; we are full of our own prejudice and bias, preconceived ideas of what things sound like. Listening is hard work, it takes a lot of concentration and to be in the moment for a modern human in his or her frantic life might seem a very odd place to be.

So, like any skills, and to develop fully this ability yourself, you need to practise. Initially, you might want to go out of your way to do this, to find time with this exercise specifically in mind. Go out on a sound-search; find a place and breathe into it. Go looking for a rich soundscape that is most likely to reward you. Eventually, you'll be able to do this anywhere you happen to be: in the garden, the park, at the bus stop, even the tube station. But, for now, find somewhere quiet, and by this I mean free of people where you won't feel too self-conscious and where other folk won't distract you. Make it a time dedicated to your ears, give yourself time when you can relax and when you might find it easy to empty your mind of all other concerns and distractions.

The rules are only one, and it sounds almost too straightforward: you simply have to *listen* – but listen in a careful and attentive way. You need to be able to listen mindfully. Most of the time we hear unconsciously. We need to be able to turn off autopilot and flick our audition into manual – to take control.

The practice of mindfulness is met with scepticism by many; 'mindfulness' is a word associated with alternative therapies and medicines and isn't, perhaps, taken seriously by the mainstream. However, when you change your reference point and realise that many of the martial arts are based on this practice as a form of focus and meditation, it does change the flavour of the word somewhat.

To mindfully listen is about paying attention in a particular purposeful way; it's about being in the moment, a neutral now, free from preconceived ideas, and leaving behind anything that might clutter or cloud your thoughts.

Get as far away from technological audio contamination as you can, mute your devices – mobile phones, pagers, anything that might invade your moment and interrupt your concentration. The aim is to let your ears and thoughts leave your head; you need to get out of yourself. To make this sensory journey you have to free your mind of any judgements. Try not to predict what you might hear before you've heard anything. Empty your head and go with what you discover. This is that 'being in the moment' again. Sit or lie down, take a moment or two to free yourself from mental clutter, take a few deep breaths, settle in, get comfy and go exploring – I often shut my eyes for this. Your floozy easy-eyes are tempted to take over and distract from your ears. This mindfulness is about letting go of all distractions even those that come from within. Just the feeling of the breeze on your skin or the sun on your face can generate sensations that trigger extraneous thoughts that can sneak in from nowhere and ambush the auditory circuits of your brain. It's hard.

You have to really focus in on a sound. Explore it, listen to it as if it's a piece of music that you're hearing for the very first time on some expensive speakers or noise-cancelling headphones. Use your ears like you might use your eyes; go listening around and then, as you might use your hands while exploring, try and pull it apart, tease out its tones and timbres. Is it a dry sound, a harmonious one, or does it have discordant qualities? I often see sounds as colours or imagine tracing a line with a pencil, but I've been told this just might be down to the way my brain works.

No matter – simply find what works best for you. Imagining the sound-making mechanism sometimes helps. I don't only mean what's making the noise. Don't just think, 'Oh, that's a cricket' or 'There's a frog'. Try and climb into the sound, get sucked into its waves, ride the pressure ridges and wallow in the troughs. Imagine the actual physical qualities of the sound-making instrument. Is it a dry chitinous exoskeleton rubbing against the same, or a reed vibrating as air is forced over it? Get to the substance of the sound.

When you've got one type of sound and you're happy that you've exhausted its possibilities, in your own time, move on. While keeping an awareness of the first, go acoustically hunting for another. Change your tuning. As we're visual monkeys, only a visual analogy seems applicable, but it's as if you're focussing your ears. Over a period of time, try and find as many sounds as possible, count them (carefully, don't let the numbers take over), jump about, revisit old sounds and listen out for changes in energy; speed and volume will impart different qualities to the sounds you are hearing and change the experience.

The goal for this kind of mindful listening is to wake up your ears, not just by getting away from and eliminating everyday distractions. It's about silencing thoughts and worries other than those focussed on the sounds you are hearing. As you practise, you will start to hear the whole landscape and its audible connections. You'll intertwine with a thriving, throbbing, humming, whirring, swishing, snapping, smacking, scratching and fluting acoustic whole; a sound view, a noisy vista, an acoustic ecology. Once you've tuned in and you've dampened your personal kinetic noise, you'll become aware of even more levels of detail.

When you start hearing things you've never noticed, you're getting there, you've woken up from your unconsciousness. To be fully acoustically aware, to be truly sound savvy, takes things to a whole new level and opens up a whole realm of possibilities when it comes to increasing the intimacy in our relationship with nature.

8

Sound School

How do we hear? Simply put, mechanical compression waves in the air are collected by our ears and turned into electrical signals which are then sent to our brain for processing. We have numerous body parts that make this possible. Our external ears (the sticky-out bits) funnel the sound waves into our ear canal until they meet the tympanic membrane, which literally translates as the ear drum. This tight membrane then flexes and sets up a chain reaction which passes the energy of the original sound waves through a series of three tiny ossicle bones in the middle ear, which then create a ripple in the fluids of the middle ear. The fluid is contained in a spiral called a cochlea (a snail) and this is lined with tiny hairs called stereocilia, which are bent by the pressure wave travelling through the fluid. This mechanical movement of the hairs is then converted to electrical impulses, which are sent to the brain and processed. That's the simplified mechanics of *how* we hear. *What* we hear, however, is a little more involved.

We often use the words 'frequency' and 'pitch' when we describe the sounds we hear, quite often without actually understanding what these words mean and, indeed, the differences.

Sound waves are pulses of sound and their frequency is the number of these waves that occur during a given amount of time – a high number of these vibrations creates a sound we might refer to as high-pitched, while a low number of sound waves gives us the opposite, a noise that is low-pitched.

We can hear sounds that fall within quite a large range of frequencies and if you've got a healthy pair of ears, it's possible that you might be able to hear sounds that fall anywhere within the 20 Hz to 20 kHz range (20 hertz and 20 kilohertz – hertz is the number per second). In laboratory conditions we can do even better than that, being able to pick up sounds as low as 12 Hz and as high as 28 kHz, but in order to hear these extremes the sounds need to be really loud, which brings us on to the subject of pitch. A shrew's high-frequency ultrasonic squeaks, or the low rumbling subsonic sounds of a herd of elephants, are sounds that are inaudible to us; they don't have a pitch that we register, and therefore cannot be described. Anyone who has used a 'bat detector' will be familiar with this. These electronic devices are simply translators of the high-frequency sounds made by bats, which, via some technological alchemy, step down these frequencies to a pitch we can hear. The sound we get out of a bat detector's speakers is not the sound that bats actually make to each other, it's simply a way to enable us to hear some of their vocalisations.

Our sense of hearing is quite a complicated one and our abilities to discern certain sound frequencies vary considerably. From the age of eight, our hearing deteriorates. There is nothing quite as frustrating as taking a mixed group of families, from toddlers to grandparents, out on a summer's walk and only those below the age of twenty being able to hear the protracted ticking whirr of a colony of cone-heads (a kind of bush cricket that has swapped the shrubs for long grass), or to be on a bat walk and have only a handful of the group able to pick up the lower-frequency social calls of tiny pipistrelles as they loop the loop above their heads.

Some people hold on to the higher ends of their hearing range into their forties and fifties. Others lose very specific mid-range frequencies; some lose ability in one ear or the other. There is even a difference between the sexes: on the whole, women have a better sense of sound. Unless you're aware of the symphony of nature sounds from an early age, there is every chance you may not realise it if your hearing has deteriorated. Sometimes it takes a group sound-sharing experience like the one just mentioned for an individual suddenly to become aware of their own shortfallings in this area.

Other factors can alter our perception of sound too, from personal health, atmospheric conditions and even the time of day or how weary we are feeling. What we hear can be a very subjective thing.

So if you're not a healthy nine-year-old girl, you may well be reading this and perhaps thinking of giving up on ever being able to use your ears to their full potential, as your best ears are the ones you left behind in your youth. However, in spite of our personal variations and limitations, there are still plenty of ways of exploring the world of sound as a medium for a closer and more fulfilling relationship with nature.

Can you improve your hearing? Well, probably not, so this and the variables mentioned above might leave you thinking that your ears are of limited use and that trying to develop a better sense of sound, especially if you are of a certain vintage or have a known compromised sound perception, is a waste of time.

But nothing could be further from the truth. In some ways, your ears are an extension of your eyes. We hear things and then we train our gaze in that direction to confirm with our eyes what our ears first told us. Earlier, we considered the noise we ourselves might make, so now, with your own personal acoustic influence on the environment muffled and under control, you are now halfway there; the rest of the process goes on between your ears and the grey matter between them.

In the same way that looking and seeing are very different processes, the same can be said for our hearing: to hear and to listen are very different things – one being passive, with sound waves tickling the hairs in your inner ear all the time, and the other being active. Your brain decides which of these internal disruptions to your inner ear to take notice of and to process. We hear with our ears but we listen with our brains. This is good news, because it means that as long as you can hear something, you can train yourself to be better at listening.

Our ears, however, fill in that large blind spot that lies behind, above and to the sides of our head. The fact that we have two ears means that we get a degree of directionality just as with our sense of sight. This is why, if you hear a fly buzzing around the room behind you right now, you don't need to clap eyes on it to know it is there; what's more, you could turn around and almost instantly place your eyes on it with considerable accuracy, thanks to the directional information provided by your ears.

At one point, we could probably all listen pretty well; after all, we were proto-wild primates – but over time we develop some pretty bad habits, as the clutter of our domesticated existence vies for dominance and relevance. Remember that moment at school when, while you were staring out of the window at the fluffy white clouds, your daydreaming was invaded by your teacher calling your name? She knew your head was not in the room and as if being caught out wasn't bad enough, she'd further humiliate you by asking you to explain what it was she had just said in front of the whole class. I'm pretty sure that wasn't just me. To become better at listening we need to recognise what it is that's stopping us doing it properly. As a sensory process it can be quite difficult to listen; we're easy to distract. To really listen we need to be able to remove some of these distractions and to develop the ability to listen through a landscape, to focus on sounds both near and far, small and big.

There is no better tutor than a wild mammal, one that is much more deeply immersed in an acoustic world than we are. To have a 'quarry' and to try and sneak up on it is the ultimate test of your field skills. Obviously, all the other factors of stalking are in play at the same time. So, to best test your slinkiness, choose an animal that is highly reliant on its auditory senses.

As you may have guessed by now, badgers have 'held my hand' through quite a lot of my life. They were my 'bears', I guess, and in being so they taught me quite a lot about the etiquette of the wild; they helped me develop a set of skills and hone the behavioural tools in my naturalist's kitbag. I've always maintained that getting close to a badger in the woods or a black rhino in an acacia thicket are pretty much the same, the only difference being the potential outcome if you get it wrong – and I have. I learned from a Swazi bushman the truth behind the saying 'you only feel the thorns on the way down'.

Badgers are very sensitive to strange sound and really helped me in this respect. With most of my badger time being a solo activity, I had only myself accountable if a misplaced rustle or sniff sent my furry brethren scattering. My childhood badger watching was a steep learning curve. I started hiding up in the trees, looking down on them, partly because I felt 'safe' up and out of the way and also I was uncertain as to how they would react if they bumped into me, but also because it seemed the best place to be, if I was to avoid being seen or more likely smelt by them. However, I quickly started to push the experience; I wanted to close the gaps in my knowledge and the physical gap between me and the badgers. Doing so made seeing what was occurring in the twilight easier and, as with any intimacy with wildlife, it allows such close observations that you start to notice really subtle things, grumbles and sighs, quiet vocalisations uttered almost under the breath, or small scars and different-coloured claws. The sorts of details that are impossible to see from a few metres above the woodland floor.

The incentive to get close was such a driving factor for me that within a few months I was coming down out of the tree. If the wind direction was favourable I would sit at the base of the tree, then I started to lie down, and slowly, by increments every night, I would position myself closer and closer, until I was within a couple of metres of their front door. I was so close to them that I could hear them getting up. Low rumblings that transcended the definition of vibrations and sound seemed to come up through the soil and to my ears via my belly and I could hear their squeaky bickering and arguments too.

I would stare into the fuzzy darkness of the hole, waiting for that first glimpse of that humbug head to bob up and test the evening air. Then they would be out, sniffing, and then, when they relaxed, they would have a good scratch. Once they'd ousted the fleas and rearranged the dust and soil in their hair they would depart.

From that first glimpse in the half-light of their sett entrance to their departure from the clearing, it was critical that I kept quiet. I became unwittingly very good at it. At this sort of proximity badgers taught me not only that you must stay absolutely corpse-still, but also that even a slightly noisy breath, swallow or the sudden onset of a grumbling stomach could alert brock to your presence and prematurely end the night's mammal watching.

How I behaved became instinctive. I was developing good 'woodland manners' and eliminating the bad. We are all familiar with learning manners: we were all once taught to keep our elbows off the table, not to speak with our mouths full or that we mustn't reach across someone else's plate. These are behavioural rules, habits that at some point in time became difficult to break. It's exactly the same when learning how to be around wildlife.

'Being quiet' is a sliding scale; it means different things to different people. Depending on the circumstances, to some being quiet simply means not saying anything but in nature this often has to be 'badger quiet'. It's a difficult lesson to learn but, with patience and a lot of practice and plenty of disappointments, you'll get there.

I often forget what this means and how vital it is to field craft. It's again so simple and obvious that it's easy to overlook. But when I have to take a party of people into a hide with no glass in its windows in order to observe nocturnal mammals, it all comes flooding back.

The moment a guest chooses an inopportune moment to scratch their arm, sniff a drip back up their nose or shift the weight from one buttock to the other, I'm reminded of my hard-won skills and lessons learned.

I've had a children's badger watch cut short by one of the kids breaking wind, the building tension and subsequent relief and excitement of children seeing their first badger proving too much for one child's bowls; a girlfriend, to whom I had promised a really close view of my badgers, got the giggles when the animal decided to try and scratch its left ear with its left rear foot, lost its balance, fell over and left us in a cloud of musky anal gland secretion and with the sight of a grey 'bottle-brush' rear end disappearing into the night – those who have shared one of these experiences with me will be familiar with my scoring system on how good-mannered and quiet they were. This, of course, is as much a reflection of my inabilities and generous non-patronising nature as it is any individual's nature-watching skills.

Beat the bunny

Rabbits and deer are also excellent teachers of the subject of absolute silence and they tend to keep slightly better hours than badgers. Very few people can sneak up on a rabbit undetected, but if you can, you've honed your skills to such a fine degree that in my world you would be regarded as a grand master of stealth.

One excellent test of patience and your wild sense and sensibilities is an activity which I thought was something only I did when I had an afternoon to kill – until, that is, I saw TV naturalist Simon King demonstrate it on camera.

It's something I call the 'long lie down'. I first stumbled upon this rewarding man-versus-lagomorph challenge when my portable hide, made from an old cardboard box, blew away.

As a boy, I would look for big boxes that I could turn upside down and hide in – I would cut holes in them to look out of and, by slowly lifting the box up on my back, I could, if I was careful, walk the whole thing around. It was quite effective, as I one day discovered, in all but windy weather. On this one particular day, I had stealthily, over a period of an hour or so, manoeuvred my box and myself across a large pasture until I got very close to a colony of rabbits. All was going well, but the wind was getting up and without warning the box lifted up and off. I was left awkwardly crouched on the grass like a tortoise stripped of its shell. I froze, the rabbits stopped grazing, heads and, more importantly, ears trained in my direction. I waited for the loud thudding, the staccato burst of a *clump-clump-clump-clump* of furry foot on turf – the rabbit equivalent of a war-time klaxon – but nothing sounded. After a few seconds, the twitching tension subsided, they flattened their ears, lowered their heads and continued about their business. I lowered my own and continued too, just this time feeling rather naked and exposed without my kitchen-appliance box to hide me. What I discovered was that I was among the rabbits and had managed to infiltrate their ever-nervous gathering.

I repeated this several times, approaching without the box, and I would find that even if the rabbits saw my coming and evacuated the open field edge, as long as I was downwind, I could simply lie down and wait; in time, they would slowly, one at a time, come loping out of the dense thickets of bramble and blackthorn, nettle and dock, to nibble on the succulent short grasses. The 'long lie down' was born. I had won my stripes, I felt almost invincible, and I had achieved that which tests the guile of even the most cunning fox. I had beaten the bunny.

Deer are similarly blessed with acoustic sensitivities and to be able to 'stalk' effectively, whether you're a photographer, a hunter or simply want to experience these animals up close, they are a worthy subject on which to test your close approach.

A word of warning here – if you are less than familiar with the larger species of deer, it might be prudent not to try for close approaches in the season of the rut. A buck or stag at this time, buoyed up by testosterone and sexual frustration, can be unpredictable and compelled to act under the influence of its hormones – it's not unheard of for them to attack those who get too close.

In recent years I've started to develop an interest in stalking deer, something I'll discuss later, but this as a skill that came quite naturally to me. A background in basic field craft and tracking is more than useful for this; it feels like something I'm meant to do. Whatever your motivations, it is an ancient skill and one that seems to satisfy a very deep, primeval connection, part of our biological evolution. Just being able to sneak up on a quarry is a pure thrill and the raw rush of clarity that fires dormant synapses and takes a neurological fast track to the core of the wild you is a key part of rewilding our experience of life.

* * *

Stalking deer has been a real eye-opener for me – learning to read the land and interpret the signs, sounds and sights, with other skills following close behind. To be sitting in the crotch of a tree branch, dropping acorns for a herd of red deer as they pass by below you, to paddle up to a half-submerged moose while it feeds on lily tubers, pretending to be a drifting log, or simply being out for an evening stalk and getting the satisfaction of being able to walk right past a solitary sika hind without her even lifting her head – these are all some of the magical experiences that I've had with these animals. This is the challenge, a rich investment in time but one which gives back a hearty feeling of deep satisfaction and belonging. It's a deep-felt vital empathy with the living world. If you can get close passage to a persecuted prey species while it is unaware that you even exist, you've pretty much disappeared; dissolved in the environment to such an extent that you've become, for moments, at least, invisible. When this is something you can achieve by design, you've nailed it.

9

The Wren and the Food Mixer . . .

LIKE MANY, I like the sound of birdsong and I like melody in my music; it makes me feel well, happy, good, contented, relaxed and a whole manner of other positive emotional accolades. However, in contrast, there are certain sounds that, for some reason, set me, and probably many others, on edge; they somehow have an impact on my behaviour that seems to be out of proportion and irrational. The discordant, internal tempest of a vacuum cleaner, the violence of a food mixer, the pointless drone of a leaf blower, chainsaws, drills – both drills used for DIY and for root canals – and the scraping of flat-footed flip-flop wearers . . . they all get to me in a way that seems hard to fathom.

There are medical conditions recently recognised which are associated with sounds. Misophonia, for example, is when a specific sound is associated with a meaning and it elicits a negative emotional response in the sufferer, while hyperacusis is a negative reaction to a sound due to that sound's specific qualities, its tonal range or frequency. While I suspect that I'm not a clinical case of either of these conditions, this does confirm that certain sounds and frequencies can have an impact on emotional states.

The symptoms, if they can be so called, manifest themselves for me in the form of a deep anxiety, a feeling of unease and being on edge. Recent studies into the qualities of some of the sounds that humans find universally disturbing or distressing, such as breaking glass, microphone feedback and the good old fingernails down a blackboard, have revealed an interesting link between the sound-processing auditory cortex of the brain and the amygdala – a part of the brain associated with emotional responses.

In addition to this, many of the sounds that trigger these neural pathways are those that contain multiple frequencies of discord which fall within a specific part of the audio spectrum between two thousand and five thousand hertz. I don't think this is an accident. Some suggest that the alarm calls of chimpanzees and gibbons have many acoustic qualities that fall within this range and that this effect, known as saccular acoustic sensitivity, explains our reactions – they are some kind of vestigial, emotional, reflex response to a long-lost ancestor.

While this is purely subjective on my part, I wouldn't mind betting that many of the distress and alarm calls of other birds and mammals contain a large portion of their sound spectrum in the same range. We also touched on this a little when we were talking about choosing clement weather conditions when embarking on a first night walk. There is a climatic influence on our emotional state: the sound of a windy day, a confusion of interfering audio chaos and background noise, means that somewhere inside we feel vulnerable; one of our main senses is effectively compromised.

At the other end of the emotional response to sound, we have positive, peaceful, soothing sounds – we have music and we have birdsong. There's a lot of debate as to whether birdsong is music, but the point is that listening to birds vocalising, proclaiming their territory, is one of the most popular ways that we engage with our auricular senses in nature. We need birdsong, even if we don't notice it on a cognisant level. Birdsong seems to punch right through to our deep subconscious, in much the same way as that food mixer. The background music of nature, the skylark overhead, the song thrush, the countless unseen warblers in the bushes, they all add a flavour to the air, they are the seasoning of the seasons, by their own rhythms they let us know where we are in time and space and when a bird is in full, unbashful song, everything feels right with the world. Your amygdala is happy, and your emotional space is a positive one.

There is probably some deeper, biological common sense at play here as well. Studies have been conducted that show that many birds listen out for the alarm calls of others. This is something known as heterospecific communication; birds (and other animals) effectively eavesdrop on the conversations of other birds and animals and act accordingly. Some learn to associate a particular sound uttered by another species with, say, the presence of a snake, a cat or some other predator, while others have been shown under experimental conditions to elicit predator-avoidance responses or distress associated with the presence of a predator even when played a recording of an unfamiliar species. This is a behavioural stimulus often exploited by birdwatchers who want to get a look at a bird sitting in dense vegetation. 'Pishing', as it's referred to, is a tried and tested abuse of a bird's amygdala.

It's possible that there are certain acoustic qualities that many alarm calls share and possible that they've evolved together as a kind of cline of similar responses which benefit all the members of a bird community. Given this, it's not such a far leap to assume that in the 'wild' we did the same. I don't believe we've stopped doing it; it's just that we've lost touch and have misplaced the meaning and the significance of these natural sounds.

If you own a parrot, or another cage bird, one way to stimulate it to take a bath is to turn on a vacuum cleaner. It seems surreal, but the sound of the Hoover contains the same maelstrom of sonic chaos as a thunderstorm with its tearing wind and ballistic raindrops drumming on branches and splashing off leaves. It's nature's way of telling the birds to get a little bit of personal hygiene in and have a moment of plumage maintenance. The birds in your garden do the same, although, to be honest, I've not tried the vacuum trick on them yet.

It's possible that I'm also doing the same, subconsciously picking up on the harmonics within the chaotic noise of air, dust and motors that remind my brain of a long-lost survival strategy. While I don't run off and take a shower the moment my wife decides to run the hoover around the house, it does put me on edge, as if I were outside experiencing a rainstorm or a gale. My inner animal feels vulnerable and has to up its game, the body on high alert; the fight-or-flight response switch has been hormonally tripped and I'm on standby.

But back to the birds: all of us, probably subconsciously, use them as a measure of the state of the world, as remote sensors. Their own highly evolved language is something we are probably all sensitive to; we use their emotional states to extend our own awareness of our vicinity.

When we hear a wren singing from the bushes at the side of the path, we are in some way reaching out with our perception, hijacking it for our own purposes. If this isn't making any sense to you, imagine what happens when the wren changes its tune: it changes how you feel. It alters a little piece of you somewhere, a different pattern of neurons and an alternative sensory pathway is triggered. The alarm call of a wren is harsh and scolding – the dissonance and discord grate on our sensibilities, they put us on edge.

Suddenly, the music of song is gone and we're back to that discordant vacuum cleaner. We're listening in to the dialogue – something has disturbed that wren. It might be nothing, but in a wild, complete ecosystem it might just as equally be something. Once upon a time, not all that long ago, when an unseen bird in the bush abruptly stopped singing, it might have meant the presence of a bear, a wolf or a snake. The distinctive qualities of this sound put part of us on high alert, and at one time, hearing and correctly interpreting these particular sounds could be the difference between life and death. We are all connected by these sounds, those we make and those we perceive, it's a web of living pressure waves being simultaneously emitted and received.

A chorus of birds singing from their territorial song perches, crickets playing a similar musical amour in the grass, frogs in the belching throws of courtship – the positive sounds of aliveness, the hubbub of life getting on with living – whether at the forefront of our minds or buried in our subconscious, these tell us that there is nothing disturbing them or their prime reason to be alive. To procreate is now priority. It's a system of positive feedback which loops each life, telling others that they're able to get on with it. However, if a threat, or even the suspicion of one, makes itself apparent, then the acoustic qualities of the environment change. Songs are stopped. Like the string section of the philharmonic orchestra all simultaneously dropping their bows, the symphony is disrupted. Sex is no longer priority any more, it's been pushed to second spot by a need to live and to survive in order to have sex another day.

It's nature's incessant repertoire that we need to aim to tune in to. Birdsong, or, for that matter, any animal song, is so much more than an audio clip or a phonetic tag in a field guide by which to make a positive identification. It's a means of embedding ourselves via our ears into nature's dialogue, of becoming a silent part of the phonic appreciation. Even the vegetation and the geology make their own mark on the acoustic signature of a place; they have a direct influence on the ambience.

Being acoustically aware is an almost immediate fix. Your ability to slide into obscurity as far as the environment and its wildlife is concerned means that this act, simple though it is, is also one of the best ways of achieving complete immersion and awareness. In most cases, to be able to hear a bird is to be able to see it.

I would say that over 90 per cent of all the birds I experience in my life I first locate with my ears, which makes me question why we call it birdwatching at all. As part of a recent job for the RSPB (Royal Society for the Protection of Birds), I was employed to find the nests of a rare bird called the ring ouzel. To find a nest of this medium-sized songbird in a relatively large-scale upland landscape, you first need to find the birds. Nature, for obvious reasons, doesn't make it easy for you to locate its pride and joy. You don't simply stumble upon a nest.

Although these are birds of open landscapes, which makes them sound easy to find, the places where they breed are often quite complex habitats, with lush forests of fern, boulder fields, fissures and crevices superimposed over a curvaceous topography of broken ground and hillsides. There can be lots of dead ground conspiring against your visual senses.

All birds have to, at some point, break with silence. Part of their daily routine is to announce themselves to the airwaves with sound; an essential part of pair bonding, territorial advertisement and a variety of other things.

When they do, if you know what you're listening out for, that is when they give away their presence. When you've located them, assuming you're listening and looking at the right time of the year, they will eventually lead you to the cup of moss, grass, roots and stems around which the activity of the birds revolves for the next three or four weeks.

So vital were my ears for this task that I can honestly say that of the hundreds of nests of this bird that I've peered into over the last few years of this fieldwork, all but a few, which I can count on the fingers of one hand, have been located because my ears found the birds first.

While sounds are omnipresent and always there for the listening, during the process when you're training your ears, it's sometimes a good start to simplify things. Eliminate some of the confusion, strip the soundscape down and consequently, by removing some of the clutter, you'll make it easier for yourself.

If you want to interpret bird, or any other animal, vocalisations then there are a few pointers on how to get the best out of your ears. For a start, sound quality is best when the air is still and cool. When there is no air movement, items such as loose clothing, hoods and straps are not being continuously buffeted around, creating a distracting noise that competes with the sounds you really want to be hearing.

If the wind speed is high enough, then it will create enough turbulence to produce its own music.

If the wind is blowing through vegetation, it sets in motion a cascade of other phonics as leaves and stems rub, rattle and vibrate, all adding to a confusion of sounds, and that is exactly what it is. With so much incidental sound being generated around you, you miss out on the subtler parts of a bird's vocalisations. It's like hearing a tune on a quiet car radio, but not being able to hear the lyrics.

Back to my ring ouzels: they have a strident monotonal trisyllabic note that carries through all but the most violent windy mountain turmoil, and if you hear this, you may be forgiven for thinking that this is the bird's song in its entirety. However, listen to the same song on a still day, and you get to hear the extra features – a quieter, complex and subtle warbling song, a reedy melody sung almost under its breath, which to me sounds like a song thrush in a glass bottle with a blanket over its head.

It is also best to listen for sound first thing in the morning. There are several reasons for this that go beyond the fact that if it's birds you're listening to, it's during this time of the day, before it's light enough to start to feed, that most species make use of the time to update their neighbours and remind them who's who and who's where. Like many of your other senses, your ears work at their best early in the day, up until mid-morning, when your body is fresh and alert; as the day wears on we get tired and our ability to concentrate wanes. The cooler the air, the louder the sound seems to be. Theoretically, sound waves should be transmitted better in higher-energy, warm air. But in reality warm air usually generates other manifestations that hinder and obstruct sound transmission.

Sound waves don't just travel from the source to you, they dissipate like ripples in a pond in all directions; those that head up, and would normally be lost to the land, dissipate into the atmosphere.

So there are lots of very good reasons for you tone hounds to be out and about early in the day. These advantages are further justified by the fact that the earlier you are out, the quieter the human world will often be; the hustle and bustle of the day's activities hasn't yet started up, meaning you'll be able to wander the acoustic landscape pretty much by yourself.

Go beyond the book

Listening to birdsong is one of the most popular ways in which humans engage with the natural world of acoustics. Birds are a very useful way of schooling yourself to be aware of the natural soundscape, as what you learn when listening to birds can be applied to any other scenario where the ability to listen and interpret the world by sound presents itself. Don't be limited to birds; while they are stimulating enough, the same techniques and approaches can be applied to almost any group of animals, from bats to bush crickets.

Wherever you are, there will be birds, if not singing then at least calling, making them an excellent and accessible subject to focus your ears on. From Burns's 'sweet warbling wood-lark' to Thomas Hardy's 'Darkling Thrush' who flung 'his soul upon the growing gloom', birds and their song have inspired people for millennia. Poets love them; and their songs are deeply embedded in our culture. They weave and thread their way through our words and music, much as the notes, emanating from a nightingale's silvered syrinx, weave through the hazel and briar.

Listening to birds and being inspired by their song is a pursuit that has always fascinated and engaged man, whether naturalist or not; we've long cherished their vocal abilities and harboured the notion that their songs contain secrets, in a language the sense of which is long lost to us.

Like Wagner's Siegfried, who drank dragon's blood in order to understand the woodbird, we as a species yearn to find significance in the song of birds and a profound connection with nature in their mystical music.

Get beyond the prose, the poetry and the imagined music and there is, of course, much more meaning and plenty of secrets to be revealed simply by the act of listening, really listening. No dragon juice is needed, just your ears and your full attention.

However, when you're learning animal sounds, especially bird songs and calls, it can seem a little daunting. Their voices and all the subtle sonics contained within their vocabularies don't translate very well into our own limited alphabet. As a consequence, they are very difficult to communicate or to describe with any accuracy and because of this they can also be quite hard to remember and learn.

When we seek to identify a bird we've seen, it's a relatively easy thing to do; after all, we live in an age of numerous high-quality field guides. It seems that, wherever you are, there is nearly always a very good printed resource that will help you unravel what it is you've just laid your eyes on. A quick glance at some half-decent graphics and you've got all you need, assuming you got a good enough look at the bird in the first place. A lesson I soon learned as a beginner was that the same can't be said for sound. You can't easily visually represent a call or song; if you're lucky, field guides, if they attempt it at all, will give you, under a limited range of subheadings (song, call, alarm call), a subbed-down handful of attempted phonetic descriptions.

The problem with these is that the sounds that are written depend very much on the interpretation of the author. Some birds have regional dialects just like humans, but an author has his or her own accent and individual way of hearing their own voice. So one person's *too-whit* is another's *kee-wick, kewick, kvik, ae'wick* or even *kivik*, to use and quote the call of the common and widespread tawny owl from a selection of field guides sitting on my shelf – and this is a bird sound that is embedded in our popular culture and one which is the go-to sound effect to use in any TV drama night scene. It's somewhat surprising to see so much variation, even if the sense of the call is fairly clear in this instance.

The same exercise repeated for the common buzzard gives me *peeioo, piiijay, pee-oo, mew, me-ooow* and *peee-jah*, and then when we start getting into birds with quite a complex sound signature we start losing the plot: *dlui-dlui, dlui', dlui', dlui', dlui, didlui; tit'tit'tit'toodle'toodle, toodle, t'loo, ti-loooi, tlootlootloo; lee, lee-lee, leeleeleeleelülu ... ee-lü, ee-lü, ee-lü; ee-lueelueelu ... tluee, tluee, tluee, vi vi vi tellellellell* ... goes the woodlark in just three of these reference books. The sounds we hear and how we convey them in our own languages are highly subjective. It's just not possible to represent the multiple vibrations and the simultaneous overlap and harmonics that can be created by the various membranes in a bird's quivering breast.

So you start to see the problem here. It can get so complex that we might be tempted to give up on our ears and revert back to our primary sense. In some cases these descriptions are a help, in so far as they give us some clue as to the various qualities of the pitch, tone and rhythm of a call or song, but in many cases they can throw us off the trail when it comes to seeking the identification of a species, and just by relying on these as references, in the same way we use the illustrations, we can be missing out on some acoustic treats by simply not allowing ourselves the time to absorb the sounds and listen to them properly.

You can sometimes identify a pop song from its beat and bass line when you hear it through a wall or issuing from a car that passes in the street; the same skill set can be applied to what it was originally intended for. If you really listen, you can begin to pick up on the nuances of the natural music and the more you engage, the more familiar it gets and the more deeply you become entangled in the dazzling and beautiful complexity of it all.

There are plenty of resources out there to help you interpret what you're hearing and of course we do live in an age of multi-media digital technologies, some of which can be extremely helpful at guiding you through the technique. There are now websites, apps and e-books that are not constrained in quite the same way as the more traditional field-guide format, and many of these now include audio. But the problem still remains: how do you represent such a diverse vocabulary in an easy-to-work format? You can't; even the best of these multi-media guides usually give the most common or frequently uttered refrains, three or four at most. While the dedicated audio apps and websites are excellent once you've become immersed and have started to build up your experience, they are nothing more than a collection of interesting but disjointed sounds and noises for a novice.

In some way, you need to build up your own internal understanding of these sounds. They have to have a context and you need to start developing an understanding of syntax, otherwise they're just that, meaningless sounds. Listening and reaching out to the sound, going to the sound in your mind's eye, is that context. This means analysing the various qualities of the sounds and thinking about how they are generated. The sense and the syntax is built up slowly over time, as you get more intimate with the environment you both share.

Birds can make noises by using many different parts of

their bodies. Understanding the sound-making mechanisms, the instruments themselves, can really help you
picture the production of the sound and even help you
describe and memorise it for future recall, all part of the
immersion process.

Snipe use physical vibrations set up over their outer tail
feathers; owls, sparrows and many others click their bills;
hummingbirds rattle their wings against their tail feathers;
and bustards stamp their feet – these are known as sonations.
However, birds normally communicate with each other using
their voice, or, more precisely, their syrinx. Named after its
resemblance in shape to a pan-pipe by the same name, this,
the primary sound-production organ of birds is located just
after the main trachea splits to form the two bronchi that lead
to the lungs. By modifying the diameter of the bronchi by
either relaxing or constricting, the passage of air flowing
either into or out of the lungs is regulated (think letting air
escape through the neck of a balloon) and these membranes
of the syrinx can be set in motion, giving a series of complex
vibrations which together create what we hear as bird noise.

The bipartite structure of the syrinx compared with the
relatively simple single tube-like arrangement of the human
larynx means that two or more separate sounds can be
produced simultaneously as the air passes over them. It is
this complex device that led Pliny the Elder, when referring
to the nightingale's song, to write: 'There is not a pipe or
instrument in the world that can produce more music than
this little bird does out of its throat.' The same sentiment can
be applied to most other bird songs. It is these different

vibrations set up by several simultaneously resonating membranes that give the depth, modulation and harmonic richness and complexity of many bird songs and why it is so difficult to impersonate them with any degree of accuracy.

Now you know the mechanics of the instrument that birds play on, you can almost start to visualise the internal workings when you sit and immerse yourself in the particulars of the sound. It's what I do every spring morning at about 4 a.m. when the blackbird in my garden fires up. As soon as those chocolate tones hit the day and wrap themselves up with the dawn, I'm following them, in my mind's eye, back to source.

I've never seen a syrinx in action with my own eyes (although there are some fascinating X-ray films of birds singing online which show the musculature contractions and contortions of the bird's entire body as it pumps out the sound), but I allow myself to feel the very physical origin of the pulse, the energy given by the bird to the virgin air, its warmed breath a vapour on the atmosphere. I imagine how the bird is handing over its very own body tissues to the music it is making. I try and get inside, to see the membranes of this magical musical instrument as they relax, tighten and pulsate with every inward and outward breath. I also think of its significance. What is it that this bird is declaring to the world? It's a woodwind instrument with an effort and intent very easy to underestimate in our own thoroughly different visual world.

When a wren sings its 740 notes a minute, it can be heard in real terms some five hundred metres away. Imagine doing that. Scale it to the relative size of the wren, and to our own body size, and you're going to have to put more than just your heart and soul into your cry if you are to be heard five miles away! Various studies have shown that the metabolic energy cost goes up by 20 per cent in a singing bird like the greater reed warbler, even more so if it's a species that has to expose itself to the cooling air by singing from a song perch. Some birds will sing for lengthy periods of time too. The 'little bit of bread and no cheese' song of the small finch, the yellowhammer, can be repeated over three thousand times a day, and what before was just a nice sound suddenly sounds exhausting – it's not a mere musical performance, this is a physical effort of Olympian standard.

When a bird sets the air around it vibrating, it is sending pulses of energy out into the environment. These waves of sound are sculpted by a bird's syrinx and these species-specific given qualities and characteristics allow us, if we get a good listen to them, to understand not only what made them, but also what their intentions are. It is this greater comprehension of each sound and the soundscape at large that we are aiming for here.

When you're listening to bird sound, there are various ways you can break down and understand what you're hearing. We've already covered what sound is and its pitch and frequency, now we can look into the qualities of natural sound and how to interpret and decipher what we hear.

The most basic characteristic of a natural sound is its tone – this is a broad-stroke term which refers to the general character of the sound, described in one of the best books on the subject of birdsong and vocalisations, *The Sound Approach*, as being the equivalent of jizz. This could be the warm, rich tones of the blackbird, the thin, high song of the dunnock, the piping of a redshank or the moan of a puffin. To the tone, you can add a bit of texture: to the description of the blackbird's song you can add the adjectives 'smooth' and 'round'; 'reedy' and 'shrill' suit the dunnock; and a 'throaty groan' and 'growl' suit the puffin. This texture is something we can describe as a sound's *timbre*. It gives the sound a spectral quality and brings it to life.

We've touched on the personal variance and the limitations of our sense of hearing already, however, and rather than get hung up on what you can't hear, let's look or listen to the sounds in a different way. Most natural sounds are not pure. One way of clearly visualising the qualities of a sound is to record it and play it back through sound-analysis software, to create something called a sonogram or, more accurately, an audiospectrogram. This is a simple, graphical way of representing sound, which plots frequency against time. It turns sounds into sights.

With time running along the horizontal axis, while the vertical axis represents the frequency, the higher on the graph, the higher the pitch of the sound. A pure, continuously sounding tone would be a single frequency represented by a straight line that would run with time from left to right. A bird call or song is very different; the sounds goes up and down, producing peaks and troughs as the frequency of the call goes up and down, changing pitch.

This may seem way off topic for a book about 'rewilding', but in this case, by stepping back briefly into the technological world, it is possible to give yourself a greater understanding of what you're perceiving. I know many people, pure birders amongst them, who simply don't get on with sonograms, but who learn to read them, and I think you'll find them a really neat and beautiful way of 'seeing' into the depths of some of the sounds you are hearing. There are plenty of examples of recordings online where you can listen to a bird vocalisation in real time and see the shape of the trace it makes. You don't even have to use your imagination. Choose the song of a bird you are likely to hear in your own garden or patch, and every time you hear it, try and visualise the sonogram. In time, you'll find it makes a lot of sense.

It might be useful to have an example of how useful sonograms can be for helping you 'read' the sound waves. Take the simple contact call of a warbler: many of the 'green warblers', while they have a signature song that almost everyone will be able to recognise as being different – a chiff-chaff, for example, sings its name, while a willow warbler gives a lovely rich descending cascade of notes – if they're not singing, but are making their contact call, then you might feel you're in trouble. How on earth are you to get to the bottom of this if the bird is saying *hweet* unhelpfully at you and when you look at your field guide, it seems that both species have a contact call which is described as *huit* or *hoo-eet*?

Have a look at the sonogram for each species and it all makes sense. The sonogram of the chiff-chaff has depth to it caused by several harmonic layers. It also has a steeper rise on the main call note, while the willow warbler emits a simple sound less in the way of harmonics and so purer in tone, but it stays on the first syllable of the call for a fraction longer, giving it more of an obvious upwards sweep. The point of the sonogram is that it can transcend our difficulties and limitations in discerning and describing what we are listening to.

I find it quite handy to try and picture these sonograms as best I can when I hear birdsong. I find it a useful way of visualising the sounds I'm hearing. Obviously, the resolution of the human ear has a limit and while we can often get the overall gist of what's going on, the reality is that the often-overlapping modulations that give a sound its depth and texture are difficult to differentiate; you can describe the overall sound but it's sometimes a bit of an ask for our ears and brains to separate the elements. However, remove some of these harmonics and you'll notice the difference, even if you can't tell what, specifically, has given it a certain quality.

As well as being a way of helping you visualise the sounds themselves, sonograms can help us understand other aspects of natural sounds which give us a scaffold of visual shapes on which to pin the unique qualities of a bird's song, a bat's chirp or a cricket's buzz.

I know many birders and naturalists who have far-from-perfect hearing for one reason or another and while I do meet those who can no longer hear clearly birds that sing or call at the very extremes of our hearing range, such as the wheeling tittering song of a goldcrest or the low 'boom' of a bittern, most can still pick up the sounds of birds which contain a more complex range of frequencies. Take the piping call of an oystercatcher: its often-repeated phrase 'my feet' is a common sound of shore and salt-marsh oystercatchers. If you were to record this song and then construct a sonogram of it, you would roughly get a pointed shape that looks a little like an inverted 'V' with the trailing edge being a little rough and less steep. You would also notice that you don't just get the one trace.

There are several of these inverted 'V's stacked above one another, as if mirroring each other. The lower one is called the 'fundamental' and it sits around about 2 kHz; those above it at 4, 6 and 8 kHz, while quieter, are still there and they very much add to the sound. Each of these layer stacks is called a harmonic and occurs at multiples of the fundamental; these give the call depth and make a sound seem richer. The opposite is if you have many traces that don't occur at regular intervals but are still stacked above each other. In time you get sounds which are scratchy and buzzy and which by the physics of sound can be described as less harmonious. Just to make things even more wonderfully complex, some birds can have harmony and disharmony going on at the same time!

If your hearing isn't very good at one of these pitches, how you hear the overall sound will be different to how an eight-year-old hears it, but you will still be picking up some of the overall properties, the *rhythm*, tone and texture of the sound. So while you may feel that if you are coming to this later on in life, you've missed out on the opportunity to hear and connect with bird (or any other wildlife, for that matter) sounds, you don't have to worry too much, you just have to listen to these natural noises your own way.

The timings within or between a bird call or song can be crucial to identification too. Sometimes it helps to think of those sonograms again. The call itself can be composed of several sounds strung together, the rate at which these phrases or notes are delivered and the *tempo*; the way these are spaced out can all be measurable by counting and timing. The difference in any of these numbers can alter the overall feel of the sound and can therefore alert those who are listening to a different species, a different context or, sometimes, an individual.

An example of this when applied is found in the telling apart of the songs of reed warbler and sedge warbler – two birds that drove me mad as a beginner.

Both are denizens of habitats that are almost impenetrable to the human eye. The most you have to go on is their music emanating forth from somewhere deep within the complex and dense reed beds and the scrub that surrounds them. To my untrained ear, it was a struggle. Tonally they can be very similar, they lack unique signature phrases or hooks that I could memorise and hang on to and, added to the fact that they also are pretty good mimics of other species of birds, including each other, I was thoroughly confounded and frustrated.

When I was told to listen to the beat, suddenly it became a breeze. The reed warbler, no matter how eccentric its improvisation, has a constant, rather plodding tempo, repeating phrases one after another, as if it's a little bored about showing you its entire repertoire. The superficially similar sedge warbler is much more sprightly and syncopated. It is the virtuoso of bird jazz. There are other differences, but for me this clinched it. Add to this the image of the bird, in case you get a snatched glimpse, and you will see that the plain brown conservative tones of the reed warbler go with the character of its song, while the more dashing and well-marked sedge looks like the sort of bird that might cut a dash with its funk. This anthropomorphism, the storytelling, the visualising, all serve to connect different regions of the brain. This is a well-known memory trick and by linking the bird's appearance, its rhythm, a visualisation of its song and a story, you are tying them all together to create a memory web with multiple references to the context in which it is performing. It's a process that speeds up the accumulation of experiential knowledge which you can fall back on and recall at will.

Many species, including birds, frogs, grasshoppers, geckos and cicadas, will alter the rhythm of their songs depending on their emotional state. When a potential mate is close by, or a rival male, they will up their game. It's a show-off or a showdown. A midnight pool of calling tree frogs in the Amazon, while noisy, is also a scenario where the males are playing a risk–benefit game. It's a trade-off. You're singing because a female could be listening in to your impressive voice, but, at the same time, your song is letting every frog-eating predator such as a false-vampire bat know exactly where you are squatting in the shallows. The hesitation and the sense of poker jeopardy are almost tangible, until, that is, a female actually arrives. Suddenly the sound-scape is turned up a notch, the night is filled with a loud and more rapid pulsation of sound, the jam is pumped, the rhythm has sped up and the mating game is on.

The same happens in most walks of life: songbirds, cicadas and crickets will have sing-offs and increase the complexity and volume of their utterances in order to outdo the competition – even spiders will engage in a faster drum solo.

I recall lying awake in a thatched lodge somewhere in the Amazon one night and hearing a strange vibration, the pulse of a hollow tapping sound, like distant woodpeckers (which, incidentally, can be told apart in some cases simply by the rhythm and tempo of their drumming). Initially, I was thinking it was beetle larvae or termites communicating with each other in their burrows within the building's timbers, but I became more interested when I noticed the pulses were changing not only in volume but in their overall vigour; something was investing more and more effort in the sound production, and also in the rate of the pulses and the speed with which each individual beat was delivered. Eventually, curiosity got the better of me and I turned on my torch to discover, above my head, a male pink-toed tarantula busy trying to get the attention of a female who was being lured out of her sock-like silken lair. The more interest she showed in him, the faster he tapped out his message with his modified front limbs called palps – a mixture of anticipated love-making versus desperation to be clear about his intentions as a prospective mate and not as a potential meal.

* * *

The best thing about being able to really listen is that you don't necessarily need to be able to 'spot the artist' and identify it from its music to appreciate and extract a meaning and significance. It is only when you need to share or communicate this information with others that this ability to identify what you've heard becomes more important. This is why learning birdsong needn't be so difficult. Don't get hung up on an inability to put a name to a face. After all, you don't need to know the piece of music or, indeed, the name of the composer, conductor or even the instruments in the orchestra to appreciate the music. However, if you do start to really listen, your natural curiosity will kick in. As you become more immersed in the listening, you'll start to notice details; nuances and then the names will start to become associated with the experience of the sound, and you will start to connect deeply with both the sound and the sound makers.

Open your sound flaps

If you look at the ear structure of any animal for which sound is the dominant sense, you will in most cases notice a sizeable flap of skin, the pinna. This external ear in these species is highly mobile and can be trained on the source from which a sound has emanated. Sets of tiny paired muscles called the auriculares allow the ears of many mammals to be drawn up and laid down as well as rotated. These highly mobile flaps act a little like ear trumpets but, being separated on either side of the head, they can be used to give accurate positioning and, just like the binocular vision of forward-facing eyes, the ears can judge the distance of the source as well. While we cannot move our ears like Stan Laurel (although some of us have a vestigial genetic musculature that allows a little bit of movement) we can all enhance their ability very simply. Just by placing your cupped hands behind your ears, you can amplify and enhance the sounds of particular interest as you face their direction. This is especially effective if it is in a noisy environment.

For example, try to discern the song of a territorial dipper singing from the top of a rock in a roaring riffle and you will struggle, even if you're looking directly at it; at best the odd note and phrase breaks through before the rest is snatched away by the singing river. But cup your hands behind your ears while continuing to focus on the bird and the song jumps right out of the water at you. In truth, while you may be catching a little more of the sound, you're not really enhancing it, you're just cutting out the extraneous ones that are coming from other directions. You're focussing the sound by catching it from a smaller area with your hands.

Electric ears

As you're beginning to realise by now, the art of listening is almost wholly tied up in the ability to focus on sounds themselves, to invest time in finding and concentrating on them. The underlying physics is the same, capturing sound waves and converting them into electrical signals. To take all you've learned so far to another level is possible, but again we have to step away for a moment from the natural instruments of detection we were born with and once again enter a world of technology. As before with the sonograms, this process allows us to gain an even greater understanding and if 'insight' isn't quite the right word, then let's call it 'inhearing'.

The technology and craft I'm talking about here is that of sound recording. While I don't intend to go into any great detail on this, it is worth a mention as it is a very effective way of gaining a greater understanding of the bioacoustics world. If you really get into recording natural sounds, then the next step is to create your own sonograms – it might be a step too far for some, but the technology is often much closer than you think. Armed with a mobile phone and a microphone you can make the next leap off into the swirl of natural phonics, a process that can help no end in the interpreting of everyday sounds.

In the same way that a camera records what we see and works very much like our own eyes, sound-recording equipment is in essence no more than an electric ear on the world. A parabolic reflector is a funnel, a large version of your outer ear, while the microphone is analogous to the middle ear, turning the sound waves into electrical signals just as your own ear does.

Admittedly, it's a bit of an investment, but by making it you can take a step further into the world of natural acoustics and sound exploration. It also allows you to record and analyse it, even collect it, so that you can make a reference collection, just as you might collect photographs of animals. In this way you can learn and develop your understanding. It might seem a bit bionic but the ability to hold on to what would otherwise be in-the-moment, ephemeral transmissions gives you a chance to repeat and decipher the moment. To me the ability that these sound recordings can give you is akin to some kind of magic and one of the few justifications for using the technological world in your own rewilding adventure.

The advantages of a parabolic reflector are that it has the ability to capture sound over a wider area and direct and focus it on the microphone in the centre. Then, by listening to it live through headphones, you are effectively achieving what you did with your cupped hands but in a way that eliminates even more of the extraneous sounds. Because of the large dish, you're also able to hear many of the lower-pitched sounds more clearly as well as some of the more subtle sounds that are easy to miss.

While we're on the subject of 'electrickery', there are many more ways to listen in on nature and sounds. You can even achieve an audio perspective on the completely secret world of creatures that operate outside our own pitch range – devices such as bat detectors can step down the high-frequency ultrasonics of bat, shrew and even various parts of an insect's acoustic world that would otherwise be well out of our range.

Standing with your eyes on a crepuscular sky, with bats streaming out of a cave entrance like a million motes of smoke, is spectacular, but somewhat two-dimensional; you're not getting the full story, as they operate in an acoustic world beyond our own. Flick the switch on a bat detector and your ears are filled with pulses and popping, smacking and ticking, a secret world of sound. Flick it off again and you are plunged back into our own silent world, incredulous and disbelieving that, right now all around you, is a soundscape outside our particular take on reality. While this is real time, you're only getting a translation of the sound, not the actual sounds that the bats make and hear.

While it's often thought of as a bit daunting, just by the investment of a little time, like so many skills, enjoying and interpreting natural sound can be made so much easier by breaking it down. First of all, find ways to enjoy the experience, this is the most important step. After that, the enthusiasm gets going and your natural curiosity, that innate ancestral 'wild' you, takes hold of the reins. Believe that you're designed to listen and work out the wild world of bio-acoustics and you will succeed. Remember, your ancestors (not that long ago in biological terms) would have made a living by knowing and interpreting the vibrational messages on the air. While you can delve as deep as you like into the physics and the science, the gadgets and the gizmos of sound, I make no apology for the technical descriptions given in these chapters. The one takeaway from all of this is that you don't need to – this is just a helping hand, a way that might connect some.

Just learn to listen and you will hear a world of previously hidden meaning in which there is a beauty and a deep connection between you and the natural world.

10

The Ciphers of Scent

Deep in the forest of Tabin wildlife reserve, Borneo, Simon stops and sniffs, comically twitching his nose. He has very youthful features for one not far off my age, not a sign of grey at his temples, skin as smooth and tight as a ten-year-old's, yet, for one with such a young appearance and equally youthful enthusiasm, he's very serious about what he does. Simon Ambi is a tracker, a naturalist who knows the forest in this part of Borneo with the same kind of well-trodden familiarity as you or I might have when we visit our local supermarket. Where I see great buttressed trees, a chaos of green, thickets of antlered ferns and palms all bound together with liana and vines, he sees food, drink and sustenance in many forms; he also reads the tracks, trails and signs of the highly secretive rainforest wildlife, but most importantly, right now he seeks building materials with which to construct a mammal trap.

He gets busy. A few silent moments of concentration, some twisting, cutting, holding loose ends between his teeth and, like a children's party entertainer making animal forms out of balloons, after some minutes Simon produces with some pride, but still a serious expression, a clever basket trap. A bit like a lobster pot, it is designed to catch an animal alive: by placing a funnel over its mouth and trapping the animal in the wicker ribcage beyond it. He baits it with some overripe banana and, using another strip of palm fibre twisted into a rope, ties it on to a horizontal branch at shoulder height. Arboreal lobsters?

'But why this one?' I asked. We'd passed many such branches, on very similar-looking trees much closer to the muddy road that we'd left way behind us now. Simon looked at me incredulously through his sensible-framed glasses decorated with beads of sweat and raindrops that had flicked up from the forest vegetation as we hacked our way through it. They slightly distracted me from the deep and seriously dark eyes behind them which were looking at me as if I was from another planet. 'Smell them.' Not a question, I didn't think, more of a statement of the obvious. He looked at me for a sign, perhaps a nod and a smile and a 'silly me, of course' kind of statement. But no. 'Smell, bad smell.' Nope, still not got it.

Then he started making big sniffing gestures, simultaneously walking around and wafting air towards his face. 'Smell, smell.' Then he started working his way along a low branch, nose to it, like a bloodhound. 'Here, here, smell it.'

I emulated his actions, and there it was, a musty, slightly unpleasant odour, a little like stale urine, an unclean toilet or a baby's nappy kind of whiff.

This, he informed me, was where a slow loris had passed by. This odour had been so obvious to him as we had walked through the complex tangled matrix of lush forest vegetation that it was what had made us stop and build and set the trap here in the first place.

Slow loris, like many mammals, live in a world of smell, a vaporous world of social-status updates: who's in a relationship with whom, whether you're new in town, are single, dominant, male or female – available in oestrus or a tail not worth chasing. It's an olfaction-based social media and in the dense, verdant tropical vegetation, in the dark of the night, it's a handy way of getting the message across without drawing the attention of too many predators. It also has a decay factor, a half-life, and so a scent mark has a temporal element; if you are a slow loris you can tell exactly how long ago the message was left. In the same way, I guess, I can tell how long ago my wife left the hallway by how strong her perfume is. Obviously, a slow loris can probably glean an awful lot more from his female's perfume fug than I can about my wife's. It's a neat way of communicating, not strange to the animal kingdom.

What fascinated me was the fact that Simon noticed it at all. After a week or so of walking the forest trails with him, I too was becoming olfactorily aware through his expert tutoring.

Every time we passed through an odiferous cloud, he would give it a species name. It might be a macaque, pig, clouded-leopard, moon rat, flowering vine, rafflesia or a fruiting tree, surrounded by a halo of fallen fruits at varying stages of decay. Via his nose, he was extracting clues as to what was around us. Hidden by the blanket of dense vegetation, other life forms were oozing and releasing chemicals into the air. Some by design, others a by-product of a biological process. Even the smell of, say, a rotting carcass, while repulsive to us, performs a purpose. The pungent putrefaction serves as an attractant to nature's clean-up task force – scavengers of carrion home in on this odour trail: like characters in a Bisto advert, mammals, birds, invertebrates all dive in to feed on, or breed on, this resource.

Whether we're aware of it or not, we walk through this world of waft all the time. But how many of us retain this knowledge of the niff, like Simon? Simon's lifetime of sniffing and questioning and re-sniffing enabled him to pass on a wealth of sensory informative short cuts to me. While there was no way I would be able to take on the nuances of his nose craft, I was becoming more tuned in to my nose than I might previously have imagined or thought possible. Training your nasal awareness, it turns out, is surprisingly similar to training any other awareness. You need to start, you need to go out of your way to smell things, to seek the scents, then slowly you start connecting, and when the source of an odourant in an environment is located, the process of linking the perfumed thing with a time and a place and, of course, with input from your other senses makes it stick. If you gain something from the process, then it sticks even faster. It is in Simon's interest, as a guide, to bag us a slow loris, it's a priority to him. Putting a visitor 'on' an animal experience is how wildlife guides make their money, how they pay their bills and how they put bread on the table. It becomes a priority and not just a hobby; it's a tool in your sensory toolkit again. Just as it would once have been. An essential way to find food and avoid trouble. Like so many other non-visual senses, it's one we've come to use less. Or so we think. Without doubt, our sense of smell is very useful to us: we use it when we eat, to test food, to make decisions about whether to put something in our mouth, or to detect how clean something is. It has the ability to trigger memories, and it has a role to play in our sex lives too, how we choose mates and as part of our courtship; it also has a role to play in our immune systems, and in our social lives – but despite all this, we don't seem to be aware of its full potential as a sense, and most of are still convinced that we're not very good with it.

How do we smell?

Like most of us, including the great Charles Darwin, you probably think your sense of smell is pretty poor. It is probably the least appreciated of the big five. Occasionally, it might make you lean a little more towards a flower, or test whether the milk in the fridge has gone off. It's a widespread misconception that as a species we've not been dealt a very good hand, and therefore our few forays into the vaporous world of olfaction and smell are pretty limited. Common belief is that as we scurried up the phylogenetic tree from a small shrew to where we are today we dropped most of our cards.

Our early ancestors had a pretty good olfactory sense, as it turns out – being primarily nocturnal, these primitive primates would have been immersed in a ripe world of odour, a black-and-white, three-dimensional landscape to which smell gave a colour. These odourants told them all they needed to know about each other and their food; they were living very much like our slow loris today. However, it was thought that as eyes became better developed and our primate ancestors moved out of the shadows and into the daylight, the sense of olfaction was replaced by the gradual ascendance of our eyes and colour vision as the primary method of perception.

It simply wasn't as useful to us any more. The evolution of bipedalism meant we left the ground below us, or, more importantly, our noses did. Nostrils were lifted away from the surfaces over which our early relatives once scampered, away from the ground that was once a sea of odour trails and clinging odourant clouds. They moved onwards and upwards, leaving what we often consider a primitive and defunct sense of smell behind. This change in locomotion and the evolution of a different exploitation of the habitats available is probably one of the reasons our senses swapped over, but the physical practicality of the limited amount of space for these sensory organs on and in our heads has been given as another. We don't have enough space in our brain case to pack in all that neurological plumbing and processing power; one sense, it seems, would have to give way to another and that is pretty much what we believe happened.

There are, however, many humans with a fine 'nose' who might disagree, many for whom the nose is a very important way of life: the wine testers and sommeliers, cork analysts (yes, they exist), perfumers, food scientists and aromatherapists are all sensory analysts of sorts that put a huge onus on their sense of smell. Are they simply highly sensitive people or is their acute sensory skill one that can be learned and enhanced by a training regime?

The way we tend to access a mammal's smelling potential is by counting the number of functional olfactory receptor genes or ORs. Recent scientific studies by the University of Tokyo have done just that and the results seem to back up what we think we already know: the best of the smellers also has the most versatile nose. The African elephant has an incredible 1,948 ORs, with mice and rats owning in the region of 1,100. Primates didn't do so well and generally have a much lower OR count, and while we have around 396, there are other primates with even fewer olfactory receptor genes. It seems that as far as some of our closest cousins are concerned we are actually pretty well endowed in the nose.

Between 2000 and 2003, a series of behavioural studies were published that looked at the bigger picture, not just the genetic evidence. They went beyond the microscopic scrutiny of our genes and took a step back, carefully considering all the factors that influence our sense of smell. They investigated the structures used in the collection and perception of smell such as the shape of the nasal cavity, taste, the brain's processing powers and language – all of which may combine to give us a better sense of smell and odour perception than previously thought.

It was discovered that if part of the brain of a rat is removed, including 80 per cent of the part that sits just above the nasal sinus called the olfactory bulb, incredibly, it doesn't appear to affect the animal's ability to perceive smell. So if the remaining 20 per cent of the olfactory bulb can still work the 1,100 OR genes, without any noticeable decline in performance, this suggests that there is no reason why a human, even with our much lower OR count, can't perform as well as a rat!

The studies have shown this to be true. We are actually very good at discerning smell and in various tests designed to see just how well we detect specific smells, it has been proved that in some instances our noses are more sensitive than those of both dogs and rats.

This experimental evidence goes a long way to suggest that we're probably in the realm of being macrosmats – that is, good smellers like dogs, cats and horses, rather than microsmats, which, previously, we were classified as. To me this just reinforces something that I've been aware of for some time, namely that we should start to use and trust our noses much more in our everyday explorations of the environment and not just our dinner plates. The nose and its associated internal software is not as shabby as you probably once thought and, with a little consideration, your sniffer could be up to snuff.

11

Smelling Your Landscape

A s a naturalist, the use of my olfactory skills is something I'm rather proud of – I can smell, or at least notice the smells of, things that many who walk with me will regularly miss. I once detected the pheromone of a female lackey moth on a walk; it was an odour unfamiliar to me and I deployed my eyes to search out the source – an investment of time and effort not many would bother to make.

A simple exercise in awareness will bring even the most urban human up to a pretty high standard very quickly (as you will see later on in the chapter). Some folk seem to have special skills but I believe we're all capable of a lot more in the smelling sense if we could be bothered to put our nose to it.

One lesson regarding this under-appreciated super-sense was taught to me in a most surreal setting.

* * *

'We'll start now, if you're ready.' A kind, soft, feminine voice, with a hint of the American Deep South, percolated through my headphones, signifying that the experiment was about to start. I was taking part in a filmed experiment, a guinea-pig participant in a scientific demonstration exploring our sensory abilities. The details of what we were doing, I was informed, would become apparent in time. For the time being, I had been told that I had to trust my production crew implicitly. I couldn't know why I was there or indeed even the nature of the experimental procedure; just knowing this would bias my responses.

Minutes before, I had entered the modern brick stark-looking building with the words 'Monell Chemical Senses Center' in gold letters above the doors. There was a clue to my immediate fate in that name, as there was in the large golden sculpted representation of a human head peering out of the wall and looking – or perhaps sniffing or listening – at me as I entered, but as to the finer details I was in the dark.

Well, actually, I wasn't, I was sitting in a six-by-six sound-proofed cupboard. There was a computer screen, with keyboard and mouse in front of me. There was no window, just a door that was now shut firmly behind me. The sucking noise it made as it closed again confirmed that if I was inclined to scream, nobody would be able to hear me. All that connected me to the world outside the room was the umbilical cable of the comfy headphones I now wore clamped over my head and through which I was about to receive my instructions. A remote camera was in place in the corner of the room to film my reactions.

The whirr of the air-conditioning unit in the ceiling ceased and the test started, a not-too-taxing series of simple questions asked by the gentle voice in my ears about various scenes and visual tests delivered to me via the computer screen. After twenty minutes or so, the session finished, and I was allowed out of the quiet cubicle for a coffee. Apparently, I was a quarter of the way through the test. The second session was very different – the experimenter's voice was ragged, impatient, rude and at times quite derogatory and insulting. If I got one of the questions wrong, I was berated. I felt stupid and horrible. The air-conditioning fan came on and I left for lunch before another two sessions in the sound booth. I was dreading another humiliation, another earbashing, but for these final two sessions the voice was calm again.

When the experiment concluded, I was finally introduced to the experimenters. Who apologised for being so horrible to me before explaining that the torment and the psychological trauma was all part of the experimental design and that they were testing my ability to pick up on highly dilute odourants in the air, at levels that can only be detected subconsciously. So what had they been doing to me? Well, the first session had been a kind of control to see how I would perform under normal conditions, the questions were easy and the questioner's attitude was a kind and compassionate one; the second test was the opposite, designed to be harder. These questions were very taxing, and the derisive voice was aimed to give me a negative experience. In addition, they had pumped into the room, while I was having my coffee, a very low concentration of an odourant – I was told it was a substance extracted from a rare pine, found only at altitude, a specific smell that I was unlikely to come into contact with in everyday life. My brain had now, like it or not, been conditioned to associate this dilute odour with a trauma. After lunch, we had another two sessions – these tests were designed to be equally taxing, the only difference being that while I was taking part in the first one, the pine odour had been pumped into the room. Guess what? The outcome was that, even though I should have performed comparably well in both tests, the one with the odour of the pine tree present put me on edge, my heart rate was elevated, my ability to concentrate and cognitively function were compromised. The pine odour had somehow become so entwined with the negative emotions I had experienced when I was first exposed to it that I had formed an olfactory memory.

It is thought that this sensitivity and ability to make such connections in our subconscious could have a bearing on all sorts of human conditions, including post-traumatic stress disorders (PTSD). The fact that you might not even be aware of the trigger for the change in your emotional estate might also explain those situations in life where, for no apparent reason, you are set on edge or get a feeling of unexplainable foreboding. A few parts per million of a volatile chemical could be all it takes.

It says 'pine fresh' on the tin. I give the aerosol a little curious squirt. It doesn't smell much like a pine forest to me, but there is something in it. Sadly, a toilet air-freshener is the most aware many of us will ever be of the natural ambience of odour. Take a walk in a real pine wood and you get a genuinely fresh smell, but what exactly is it? Why does pine smell like it does? Which bit of the pine is smelly, from where on the plant does it emanate?

As you pass through any environment, you are not just pushing through or breathing in air; that ratio, learned at school and quickly forgotten, of 78.09 per cent nitrogen, 20.95 per cent oxygen, 0.93 per cent argon and 0.04 per cent carbon dioxide, with a little water vapour thrown in, is, thankfully, an over-simplification of the reality. The stuff that creates memories, fires the emotional part of your brain, stirs up deep and evocative thoughts and gives you millions of clues about what is all around you every time you inhale is also mixed into the sterile definition of the air you breath. What's more, these particles of other chemicals are bouncing around in varying concentrations, some released steadily by biological processes occurring all around us, such as the leaf litter, the soil. Others are more discreet, they waft up in clouds, concoctions with purpose: the scent of sex, anti-predatory strategies of plants and animals, a decomposing thing unseen in the bushes, a deceptive lure, an advertisement from flower to insect. The very air we breathe is a communication channel for so many living processes that it would be remiss of someone on a self-rewilding journey to ignore it. We live in cloud of smell, an invisible soup of volatile chemical influences.

We've seen that we're capable of smelling so much more

than we might think, and it has been proven that by consciously and mindfully drinking in particular smells you can improve and train your abilities. In the same way, we've demonstrated that, just as by being hyper-aware with your sight and hearing you can really open up your sensory horizons, the same can be done with scent.

Natural scent, odour, malodour, odourant, smell, call it what you like, is produced by everything. As with other forms of non-visual sensory stimulations that don't engage with our primary sense of sight, I end up using my inner eye to explain it. This does follow some kind of logic if you think about what odour actually is. Although the volatile chemical compounds that we register as odourants are invisible, they are still physical entities, they are particles, albeit very small molecular ones. It is these that make their way into your nose and tickle the sensitive hairs in your olfactory bulb. I find it helpful to imagine them and how they get there.

You might be walking along the edge of a field, it's early on a December morning, it's cool and damp, the air is still. It's a perfect day for smelling a smell. Then you get one. A sudden whiff of something powerfully musty, you've penetrated a cloud of it, a split second and you've passed through it. You stop and retrace your steps backwards and there it is again, it's a finite and physical thing, this cloud of odour; it has edges.

Combine a few of your other skills and you might find more clues as to the nature of the smell: a phantom trail in the dewy grass, a distinctive footprint in the mud, maybe some displaced vegetation or a hair snagged on the wire fence under which an animal has passed. Get down on your hands and knees and the smell will get stronger. It's the pungent aroma of a fox. Most of us have smelt it before, whether you know it or not. The problem is, I have no words to help you out here. Fox smells like fox – you need to have first-hand experience of a fox or for the knowledge to have been passed down to you to know for sure.

I know the smell of fox because I've been lucky enough to have smelt one. The first time I unequivocally smelt fox was one of the many seminal reference points that I had growing up in a rural location. I remember it clearly. I was helping my dad out in the garden. It was while I was on my hands and knees pulling up fistfuls of groundsel and shepherd's purse. Tearing them up by the roots, clearing the potato patch ahead of my dad's systematic clod-turning fork.

This was nothing special for me, it was a regular, routine situation. I was probably complaining at the time, too, as this would have been a chore and I was probably bribed by the promise of a little pocket money to blow on a paper bag of sweets in the village later.

As we cleared and turned the vegetable plot, my dad would point out anything of interest while we worked; a clever tactic aimed at holding my interest in the task. Most usually this would be specimens of earthworm which were plucked from the freshly turned sod and placed in his bait box for fishing. But sometimes other creatures not destined to be dangled in front of a trout. There was plenty to hold my typically short attention span in the dark, fertile tilth.

Myriads of creatures scurried from the light – subterranean centipedes were always a favourite: like marmalade-coloured string they would ripple their legs in reverse to back up into the loam from where they had been woken. There were turgid and pale cockchafer grubs and the grumpy, dry toads under the old scaffold planks Dad used as tread boards. But on this occasion, there was something tangy in the air, something of interest but not in my usual visual dimension. There was something that made my nose wrinkle. 'What is that horrible smell?' I asked.

'Fox,' Dad answered. How did he know it was a fox? I guess he had the same information passed down to him by an elder family member too, the way knowledge creeps down through the generations. This was a 'boy's own' kind of knowledge. The sort of thing nobody would teach you at school.

Dad went on to explain that a fox had probably been mooching around the garden last night, his fancy tickled by the smell of our free-range ducks and chickens. I felt the prickle of excitement. Foxes were (and still are) exciting to me.

Just over the field was an old sandpit quarry, now referred to simply as 'the dump'. Here generations of villagers and the farmer had discarded their junk. Occasionally, I would catch a fox here, sunbathing on the roof of an old rusted-out Wolseley that slowly senesced amongst the bramble and elder, a machine returning to its fundamental elements. The fox perfectly blended in with the roof of the old car and it was only that unerring sense that something was watching me that first made me look up. Then my eyes met Reynard's;

I was looking through amber windows straight into the un-tamed spirit of the land. The countryside surrounding me might have looked tame and cow-towed to the hand of man, but when I first looked into the furiously indignant eyes of a fox, I felt it: wild was very much alive in an animal of survival; a creature of wisdom, cunning and resourcefulness living amongst us. So to think that that rusty streak of wild was sliding around just a few metres from my bedroom window while I slept thrilled me; it was as if I was right all along: magic didn't just exist in fairytales, it was here all the while, unseen – but not, as it happens, unsmelt.

From this moment on, the smell of fox has stuck with me. It has cloyed to my working senses and when I pick up that tang on a walk, I immediately look around for further clues that back up the nose and usually I find them. I can imagine that shadow sliding around field margins and crossing the path where I now stand, the fox's trot hidden by the high grasses and herbs, his odour floating on mist and dew, pulled by an invisible thread into thickets that would hide the fox from our primary senses – however, with a sniff and a memory, he's busted.

If I drop down to fox level and get closer to the scent, it gets stronger, and, to mix my sensory metaphors, it gets louder in my nose. That is because I'm getting closer to the source. So what is the source? It's rarely the fox directly as he's long gone, although at this time of the year sometimes you can pick out a sweeter smell, if your paths have only just crossed. It's produced by a gland on the top side of the tail; it's a gland that wafts its volatile secretion directly into the air. The source of *eau de Reynard*, familiar to dog owners, is known as the violet gland, a name given to it because of the odour's heavily poetically licensed resemblance to the smell of these flowers.

The odiferous signature of fox – well, at least the predominant smell I can pick up – is a mixture of things, but mainly it is a little puff of urine, a sprinkle here and a dribble there. Just as our slow loris smears its world and walkways with urine, so does the fox. It's a habit that is strange, if not unhygienic, to us but it's essential to the social and sexual life of many mammals and it is one familiar to lead-tugging dog walkers. The messages in these vaporous communications are untold, but at least some of the conversation can be heard if we pay attention to smelling it. The cloud of fox vapour tells me he passed this way – and it usually is a he, females have a smell too, but nothing quite compares to the strength of odour in male fox urine.

It helps to think of it as any perfume. It's a dab somewhere on the vegetation, on a tuft or tussock, sometimes added to faeces, but it is a finite quantity and, from the moment it's deposited, volatile molecules of smell detach themselves and are wafted away by the breeze. The conditions that these scent marks exist in very much determine how long they last and how strong they are.

The still air on a morning walk means that the smell of a fox is strong, it's probably only a matter of hours old, so it's fresh; there are still plenty of volatiles leaving the original deposit, but the fact that the air is still cool means not only that it has not started to rapidly dry out and evaporate, but also that the other air particles they are mingling with have not started to liven up too much. That cloud of scent is still a cloud because the air has accumulated all of the particles that have torn themselves away from the spot of urine; each one hasn't got very far in its journey or random dispersal.

As the sun rises, the air warms, resulting in the molecules of gas picking up more energy. When they do this, they start to dance around more; they bounce around, spreading out, colliding with each other, and this has the effect of rapidly shuffling up the scent molecules and speeding their distribution and therefore our ability to detect them. Most saturated and humid air is pretty good at holding on to smells too – the niff of slow loris, tapir or macaque clings to the air, while the smell of decay and flowers always seems to be more concentrated in the tropics and this is probably the reason why: the wetter it is, the slower these volatiles evaporate and disperse.

This lesson learned from the fox is one that applies to all of the smells we detect. The process is twofold. As with any sensory awareness, we need first to notice the smell. This requires a deliberate effort to go out of your way to smell something in the first place. It's an exercise that is familiar by now, it's how we've trained and exercised our sight and our hearing. But, secondly, it is through giving the sensation a place in your thoughts – building a registry of connections that somewhere in your existence, you've tasted and smelt – and sometimes both at the same time (as the two senses are very close to one another and some would argue are even partly the same thing) – these ingredients separately. You've laid down a reference point, you've got some kind of memory.

A bottle of desert rain

So to become a more complete human, one that is fully installed in the unbridled potential of your own body, you simply need to build more reference points, and gain more experiences via your nose. This can start right now.

Vegetation is probably the primary source of natural, environmental redolence, and it is probably more important to us in recognising and gaining a sense of place than we at first realise. I want to take you away from the bottled ambience of pine-fresh room deodorants and their brutal assault on the senses to a world of smells, many of which are very difficult to bottle, except one.

On the rare occasion that it rains in the Sonoran desert, the air fills with what the locals might refer to as the smell of rain; it is the smell that accompanies the suppression of dust, a clearing of the air and usually an explosion of animal and plant activity. It's a smell that triggers exaltation among the desert's human and non-human residents, which rejoice in the abundance of this usually scarce life giver.

To find the source of this smell is difficult and it eluded me for years. During the rain it permeated all and when it was dry, it was lost in the searing heat, where the air seems to almost burn the interior of your sinuses into a parched papery submission. I had assumed that the smell was atmospheric, that it was the smell of ozone, that rare triad of oxygen molecules often generated by lightning strikes. I had been in many thunderstorms in my time but I had never smelt anything like this before, outside the desert. To be honest, I hadn't really started waking up to my olfactory perception. At this point I was nose blind; I could smell the coffee but I was far from awake to its potential.

It wasn't until I had a chance encounter on a desert trail not far from Tucson, Arizona that the mystery was solved. I was out walking early one morning, trying to get the most out of the day before the searing heat of the sun made being anywhere out of a pool almost unbearable, when I bumped into a Tohono O'odham man, out and about doing the same thing. When the conversation turned to the creatures that I had been seeing on my hike, I very quickly realised that I was in the presence of someone born and bred in the desert who had a deep love and connection with all it contained. It was during our informal chat that the question of this smell came up, as it had rained a few days previously.

His answer to me was to walk a short way off the trail and gently grasp a branch of a spindly bush and rub a few of the sparse tough little leaves between his forefinger and thumb. He gestured for me to do the same and then to sniff my fingers and there it was, right under my nose, the smell of the desert in a deluge. Locked deep within the cells of the greasewood or creosote bush, imprisoned within the waxy leaves in a cocktail of aromatic oils. These oils fulfil many tasks: to make the foliage unpalatable and to reduce predation on the leaves, to suppress other plant growth and to guard that most precious commodity in the desert environment, water. Usually these same oils are held captive but a good soaking of rain, or the friction of a thumb and forefinger, and they are liberated for all to smell.

I must confess that to this day I have a little sprig of creosote bush leaves in a glass vial. When I want to be spontaneously transported the five thousand miles back to Sonora, to sit beneath the towering Saguaro cactus, hear the dry buzz of a rattler's tail or the scolding of a cactus wren, all I have to do is pop the lid, shut my eyes and gently inhale.

This phenomenon is not restricted to the deserts of the American Southwest. You can smell the rain almost anywhere that experiences heat combined with sporadic rain. It's not the same as the distinctive aroma of the Sonoran desert, but it does have enough of a similarity to make you suspect it shares a common chemistry.

When rain has fallen on a landscape that has undergone a period of prolonged drought, something happens. I've smelt the rain like this in the red centre of Australia, in the Serengeti and in a Tesco car park in the middle of a summer thunderstorm – places where there isn't a single scruffy greasebush to be seen.

The smell of a landscape is often released by rain, it's such a noticeable phenomenon that it even has its own word, *petrichor* – derived from the Greek words for stone and fluid, or blood. While this has been wholly attributed by some to the production of ozone during lightning storms, the variability of the smell suggests otherwise. It's neither the water nor the ozone, both of which are fixed in their structure and molecular make-up, but a combination of molecules as unique to the landscape as its geology and the species on which the rain falls.

Different intensities and odours suggest that the smell is of a place, it is in effect the essence of the landscape itself. Something in the living landscape lifts away from its earthly tethers and, for a limited time, cavorts and gambols, albeit for a brief moment, in another gaseous element, before shooting straight up your nose.

When you smell the *petrichor*, you're smelling not the rain but the landscape itself. Just like the distinctive and unique oils in the greasebush found in the American Midwest give the desert its post-rain smell, other plants' oils also contain and utilise varying combinations of similar oily, volatile and aromatic compounds such as terpenes, limonene, camphor, methanol and 2-Undecanone, to name a few. During rain, these oils, many of which drip down over the course of the life of a plant and find their way into the soil, along with a substance called geosmin (which literally translates as earth smell), a waste product of soil-living bacteria, combine to form distinctive-smelling aerosols.

Bubbles of air released from the saturated soil percolate upwards and in doing so initiate the process of picking up and liberating these various oils and metabolic acids into the air – the result is an environmental odour cloud, unique to the species of bacteria and greenery found there.

You don't have to wait for the rains to appreciate these smells. If I had crushed the creosote bush leaves first myself as part of a more complete exploration of my desert walk, I would have identified the smell of the desert first-hand without the need of some local wisdom. Nowadays, and as a direct result of my Tohono O'odham lesson, I like to create my own sensory wisdom and enrich my own experiences.

As I walk, I've now developed a habit of plucking at the leaves of plants as I pass them on the trail and rolling them between my thumb and fingers to release the chemistry within. By investing time in my sense of smell like this I create many extrasensory connections. Some are conscious while undoubtedly others will be filed away in my sensory memory banks. At one level it is a simple pleasure, like enjoying a sunset is for your eyes, or listening to a nightingale is music to your ears, but drift a little further into the world of smell and you'll discover a world of practical applications.

I need to give a word of warning here: at this point it is preferable to apply a little bit of local information and that other inward wit, common sense. At home, I don't make a habit of grabbing, say, the leaf of a stinging nettle or a bramble – as I know, or, more critically, I've learned, that these plants are not as innocuous as they look and are capable of fighting back. The day I grasped at one of the soft heart-shaped leaves of a stinging bush in Western Australia was the first in a week of agony and sleeplessness, with a fire on the skin that seemed to be revived every time I sweated or took a shower. I've repeated the mistake with poison ivy and poison sumac in America as well. It's not always chemical either: while you are unlikely to grab something that looks thorny or harsh to hold, some grasses and ferns have stems that contain silica and can cut like a razor when drawn between the fingers. I guess there is no faster way of learning what not to touch in the future!

What you learn, whether via an odour, an unpleasant chemical reaction or a good old-fashioned pricking, is that in the process of crushing leaves you will develop a deeper sensory insight into the environment. By using your sense of smell as part of your routine explorations, you begin to train your nose to notice. When you start, you'll be surprised what you tune in to. It's such a part of the ambience of an environment, and to realise the scented signature of a habitat gives you as much a sense of place as any other sensory stimulus and sometimes even more.

12

Tree Spotting

ECENTLY, I was taking a group of college pupils out on a bit of a tree-identification ramble. A simple ramble where I was running through some of the basic features to look out for in order to put a name to some of our more common woodland and hedgerow trees –a skill that, while being rather fundamental for all sorts of practical reasons, develops yet another reassuring familiarity and connection with our environment. You may never need to make clogs or a piling for a pier, or collect nuts or make jam, but to know that alder is the best rot-proof wood, that hazel is the tastiest nut, and that wild damsons are the ingredient for a tasty wild jam or gin, all helps with giving us yet another dimension to a walk and another link to our heritage. A familiarity that breeds intimacy, from which a love and appreciation is born – like all good relationships. Knowing your trees gives you clues to the geology beneath your feet, the historic use of the land and the other kinds of plants and animals that you might find in the vicinity. While exploring the trees as we passed them, up came a very poignant point.

I had in front of me an elder tree. It was just coming into leaf, its purple leaves pushing bud scales aside. Like so many of their demographic (a recent survey came to the conclusion that 98 per cent of us Brits cannot identify five common tree species), these boys didn't know their oaks from their aspens. This elder was just as strange to them. So I began to explain and pick my way through the distinctive features.

It was, I thought, going well. I had just demonstrated the pith-filled stems which could be fashioned into whistles and pea shooters; confident that I now had their attention, I carried on and then I took a leaf and crushed it and explained how easy it was to identify the tree by smell alone. I had lost them. The fact that its leaves smell sour and acrid when bruised didn't help. 'It smells of wee,' someone added helpfully.

Why is smelling flowers acceptable but leaning in to smell a leaf for the most part is a bit weird? After all, those with a culinary bent will do exactly this with herbs and spices, as a way of assessing their use and suitability in creating a flavoursome dish.

With the smell of the elder firmly in nostril, it got me thinking. Just the smell of this tree reminds me of so many things – wine and foraging, badgers, garden fires, rainy summer days, moth trapping on the roof of my dad's garage. The pungent elder leaf somehow had the power to link so many aspects of my life in one sniff, a web of seemingly disjointed and varied experiences all connected by the smell or smells of a single tree.

The elder is the tree of three, three smells, that is – the light, dusty and gentle smell of the flowers in the spring, the rich, sweet, sugary, boozy one of the berries in late summer, and, of course, the ubiquitous smell of the leaves. Both of the former represent two natural events that occur as discrete moments in the seasonal cycle: clear identification and demarcation of the seasons. You cannot fill bags with the puffy umbels or tubs with the purple finger-staining berries at any other time of the year – they identify a time, and the memory that these smells evoke gives you the place. It's a deep connection which can be as vivid and clear a reference point as seasonal, very commercial dates in the diary. When you synchronise with these rhythms, you're connecting in a very intimate, practical and human way.

The stair rods of summer rain seem to agitate the elder in the same way as when you brush past it. Walking along a field margin in a heavy downpour, you'll smell the elder foliage more than any other. It's a tree with presence, and it lets you know it is there, even if you're not looking for it or if it has no flowers or berries to offer you at the time. It was also the only tree available to shelter under on the way back from the beach on our annual family summer holiday; another cluster of reference points for that smell, another memory lodged: the soft, dusty bark adorned with the bright orange lichen and the soft, downy flaps of the jelly ear fungus.

It was an elder that grew behind the single-storey corrugated tin garage, previously a stable building. No doubt deposited there as a seed in the purple dropping of a bird that had perched on the wiggly tin roof edge many years ago. It had grown close to the wall, not big and rooty enough to cause any structural damage, and therefore it had managed to avoid the death warrant of the chainsaw. Because it was growing in the shady waste ground, the no-man's-land of the garden, it had been left to do its own thing. It was a handy ladder for me to climb up into the roof; its soft, innocuous foliage and multiple branches made it easy to push up through and here was a perfect place to hide out with a clear vista of the garden, and a good spot to place my homemade moth trap – a device akin to a lobster pot for trapping live moths, the bait being an illuminated light bulb. Here under the topmost spread of the elder I would sit with pots, specimen tubes and a copy of Skinner's *Moths of the British Isles* and pore through its pages. I even found the cryptic caterpillar of the swallow-tail moth feeding on its leaves.

The stems of an elder can be used as a pipe to blow life into the embers of a fire. The garden 'burn up', while frowned upon by the ecological gardeners of today, was a very much looked-forward-to event when I was growing up. The connection between the smells of the bonfire's wood smoke, of marshmallows and elder, is one derived from the old use of the tree, and one that, possibly, is part of the etymology of the name. 'Elder' as a word may well be derived from the Anglo-Saxon word *aeld*, meaning fire.

All of these experiences and sensory adventures can be conjured up from somewhere deep in my head by just the whiff of those crumpled leaves in my fingers. This is the secret power of smell. More than any other sense, that of olfaction is able to release memories and alter our emotional state. This in itself is a reason to value our senses and the organisms that provide them. These deep-set, multi-sensory connections also have a practical use: they are an *aide-mémoire* in helping recall natural foodstuffs and plants with handy properties, many of which might still be of value today. However, these skills are slowly being erased and eroded by our less nature-centric commodities and cultures.

The reason for the poignancy of smell and its uncanny ability to reform long-forgotten and deeply buried memories is this: smells are received by the receptors in our nasal passages, the nervous impulses they generate then pass through the olfactory bulb and then straight into the limbic system – the part of the brain that has been shown to be responsible for emotion and long-term memory. It is a short journey directly to a primitive part of the brain, which some call the paleomammalian system – something else we have our early nocturnal tree-living ancestors to thank for.

The ability of a simple smell to trigger a trip down memory lane or even an emotional state is a very powerful tool for those seeking to immerse themselves more fully in the natural world. The power of this ability of smell to alter our emotional state pervades most of our consumer life, so why not also nature, in its raw and original sensory form?

Of course, we can't end this sensory journey initiated by a tree's leaves without asking why? Why do elder leaves smell so strongly? The flowers have an odour because, like all flowers, they are designed to be attractive to pollinating insects, the fruit smells to maybe attract mammals such as us to disperse the berries and therefore the seeds with them, the foliage smells because . . .

Well, it is less than tasty, and while many species will consume the leaves, it is a figure much lower than, say, the leaves of the hawthorn. Toxic substances such as the alkaloid sambucine and a cyanogenic (bonded with cyanide) compound, sambunigrin, are contained within and it's because of these that it has many traditional uses as an insect deterrent, either as a potion to be topically applied to stave off the attentions of midge, mosquito and horsefly, or simply as a bunch hung from the harnesses of working horses, by the stable door, or planted around the back of the house to deter flies from entering the kitchen and the pantry.

Smell and the forager

Smell has a direct and very pragmatic application too. Just as appearance and sound can be useful in identifying anything from a redstart to a red deer, smell can be used in the identification of many other less-than-extravert species.

Carbolic, sulphuric, milky, aniseed, goaty, rotten meat, as well as coal gas, garlic, nail-varnish remover, public swimming pools, potato peelings, hen houses, Russian leather, rubber, even the smell of sperm and harlots have all been used to describe and help identify different species of mushroom.

Many wild-food foragers love the idea of mushrooms, and as far as sustainable meals and wild foods go they tick many boxes. However, in a world where all the difficult decisions are made for us, the supermarket shelf seems a safer bet. In the mushroom world, the difference between dinner and fatal poison comes down to how a fungus smells. As with much foraging, there is nothing like the focus of possible death or a very unpleasant bout of stomach cramps to make you pay attention to the details.

I do encourage foraging for this reason; it gives you a reason to get it right. In some ways it reminds me of my Alaskan bear, we all need a bit of jeopardy in our lives and if we can cheat it, or develop a skill that allows us to work through such challenges, it brings with it a genuine sense of satisfaction that is hard to beat.

Incidentally, some of the best eating mushrooms to be found in the wild are members of the agaric family, but it contains quite a few that look very similar to each other, and that includes some poisonous ones. You have to be able to tell them apart if you're going to consume them. How they smell, combined with other features – size, shape, habitat, bruising and growth form – can help you make decisions with confidence. In general, I tend to use smell last; if they tick all the other boxes and smell of aniseed, or smell mushroomy, then I'm pretty sure I'm safe, but if they smell of soap, chemicals or a bit inky, then I need to go back and check the details. The chances are I've made an error and I'll not be taking a bite.

13

A Question of Taste

You MIGHT think tasting your way around the natural world is limited to edibles, but you'd be wrong. While I'm certainly not suggesting that you go around licking rocks, nibbling on tree trunks or taking bites out of birds, bugs and other beasts at will, the idea of using taste as a way of gleaning information about the world isn't completely out of bounds.

If we're going to step back and away from our self-inflicted taboos and socially inflicted pressures and simply look at the human body, as an organ of exploration and in order to be able to take stock of its raw abilities, then we have to notice that we are all fairly well adorned with sensory receptors in our mouths. If we didn't at least pay some kind of heed to the diverse collection of taste receptors, chemoreceptors, mechanoreceptors and the thermoreceptors found here then we could be accused of being incompletist.

Admittedly, our sense of taste takes a bit of a back seat when it comes to the other senses in the field, but it does have its applications, although they are rather specific ones. One of the issues with using your sense of taste is that it is usually a precursor to consumption; most things that touch the tongue are on their way down already. It's a doorway into our insides. We are rightfully nervous about the risk of poisoning and infection and therefore we have in place a strict set of admission rules to our guts. Some of these are fairly basic common sense, others stem from what is culturally accepted. You wouldn't eat rotten food, would you? Well, you almost certainly do – fermentation is controlled decomposition; when you eat yoghurt or miso you are consuming products that are, in effect, rotting. How about a fish head, putrefying in its own juices for weeks underground – well, in Alaska the heads of salmon are treated in just this way before being mashed up and eaten – a delicacy known by the matter-of-fact name 'stinkheads'? So you see, what we let into our bodies and what we don't isn't quite as clear-cut as you might think. One of the quintessential rules of exploring is to be open to new experiences, to not be judgemental and to keep an open mind.

I'm not concerned about actual consumption here, and that aspect of gustatory exploration is of more relevance to those seeking wild foods and foraging. However, part of the process, the first part, that is, when a substance is taken into the mouth and tested, is a legitimate way of exploring, and while not one commonly used, it does have its place.

How we taste is not a simple thing to describe. Most commonly we think of taste in relation to taste buds – there are some ten thousand of these in the mouth, most of which are distributed on and around the tongue although there are others spread around the buccal cavity and throat. Each one of these taste buds sits around a hundred taste receptors which, between them, are known to register the five different taste sensations – saltiness, bitterness, sweetness, sourness and meatiness, also called *umami*. However, taste is just one of the many factors in a slightly more difficult to pin down sensation that we refer to as flavour.

Naturalists using taste as a tool to explore the world isn't anything new. For example, there have been famous historical characters who have made a habit of tasting pretty much anything they can get their hands on. These proponents of the pursuit of zoophagy are now viewed as eccentric, but the father-and-son duo William and Frank Buckland were fine practitioners of what they thought of as a legitimate science.

During the nineteenth-century heyday of zoophagy this method of exploring the natural world with the taste buds was merely a form of adventure into the world of curious questions; its proponents were seeking experiences using the means that their own evolution had provided – in some respects, you could say they were rewilding, revisiting processes that undoubtedly had taken place at some point in our history.

While it may seem that these sorts of exploits should remain confined to history, Frank Buckland wasn't seen in his time or by his contemporaries as being all that eccentric, he was merely embracing the age of exploration and enlightenment in yet another way. Global exploration was revealing a world of bizarre and hitherto unimaginable beasts, and human cultures with strangely different diets; the Bucklands were very much at the forefront of exploring the potential of other food types and assessing their qualities and potential as mainstream food sources as part of the acclimatisation movement.

However, the limits of zoophagy are really a subjective matter. After all, the only thing stopping us finding squirrel or crickets in the supermarket aisles is social conditioning and taboo. For the true and open purpose of this book, and for keeping the Bucklands' sense of exploration alive, we will briefly touch on it, but within the context that we are utilising our taste buds in ways that are not associated with consumption of food and if used at all, are more about precursory explorations of our world, in particular in areas where our other senses can't help much.

The curious case of the dancing slugs

I've always liked slugs. It's something to do with my natural propensity to stick up for life's uglies, its evolutionary underdogs and the creatures people like to hate. Slugs come pretty high on that last list, thanks to the predilection of a mere handful of species to ravage our radishes, predate our pansies and desecrate our dahlias.

Where I live, the moorlands of southwest England seem to almost ooze with shell-less molluscs from spring right through to autumn. One species in particular, the great black slug, is exceptionally prolific and noticeable. During a five-minute slow walk along one of Dartmoor's narrow tracks, I counted 188 of the things, each seemingly a good dollop of animal protein for any species fond of *escargot*.

However, one thing struck me: these animals seem to be flaunting themselves in a very un-slug-like way and in the twenty plus years I've been living here I've never witnessed any animal eating one, save for one brave or naive mistle thrush that seemed to flick one around for a few minutes like an unwanted dumpling covered in glue.

This, and the fact that they are slow-moving, obvious to all and big, has led to the development of a personal pet theory. Pick one up and poke it around a bit as if your finger was the sharp end of a curious bird-like predator and they retract, their previously impressive length of over fifteen centimetres snaps back like a piece of living elastic to a large gooey lozenge, with the appearance and texture of an oversized wine gum.

In doing so, the thickened mantle skin, something analogous to a medieval leather shield, is presented to the attacker, then, if the attack persists, the slug starts to 'dance'. No fancy footwork, that's difficult if you've only got one foot, but more of a slow, lubricious wiggle.

My curious nature wanted to know why. There are a lot of questions posed by this animal. Why does nothing appear to eat them? Why do they seem so bold and brazen? Why do they do the little slow and sensuous wiggle dance? They are toxic, which explains all except the dance. My theory is that the dance is actually a physical way to distribute the super-viscous slime across the surface of its body. After all, they can't spread it around like we would a body lotion, they've got no hands.

I had sat on this theory for some time, until I happened to find one of these slugs while out with a group of clients who had joined me for a moorland ramble, and I went through the story explained above. I got to the last bit and it seemed that my audience wanted more from me. I would not advise anyone to do this, but since I didn't really want to leave them hanging and I wanted to make my point, plus the fact that this question had bothered me for some time, I licked the slug.

It was not a long-drawn-out lick, more of a quick and cautious one. It tasted nasty – a taste difficult to describe but a bitter chemical-like flavour. It was certainly distasteful and immediately I could put some of my initial questions to bed. The chemical constitution of my slug's slime was almost certainly a big part of its courage, its apparent audacity to be out in the open when other invertebrates were tucked up and hiding, awaiting the cover of darkness. It certainly wouldn't be a nice meal. I carried on with the nature walk, occasionally picking the odd ball of slug gloop from my lips while somewhat contentedly mulling over the fact that my theory was shaping up nicely.

I stopped and was about to point out a spider on a gorse bush when I noticed I couldn't feel my lips quite as well as I could a few moments before, and my tongue had gone tingly and numb. It turns out these slugs have some kind of chemical that is an anaesthetic in its slime too.

While there are a few assumptions made about how any prospective predator may perceive the taste of the slug, via my own taste-test experience we can certainly begin the investigation. Using one's own taste receptors to check for noxious compounds might seem like an accident waiting to happen; after all, one is ingesting potential poisons. In the realms of chemical defence mechanisms there is probably little risk of ingesting anything that will give us a serious reaction, especially if we carry out a few extra precautions. To carry out an effective sensory experiment one doesn't need to swallow; usually just a dab on the tongue will provide all the information we could possibly want, then we can spit out the rest and rinse our mouth out with extra water if necessary.

Though I am describing a situation that I by no means advise anyone to repeat, what I aim to illustrate is the role of taste in our natural lives. To investigate some of the elements of the natural world like this might not be to everyone's taste or, indeed, inclination, but if the definition of adventure is to push ourselves to experience new things, then surely the relatively unexplored realm of tasting nature matches this more than adequately. It's a very natural instinct and a vital part of the learning process; up to the age of about eight months and before your other senses developed, your mouth and the millions of sensors it contains would have been your primary sense organ.

To me the world is full of taste sensations or flavours that can help us make sense of other observations. My mysterious slug was just one such moment.

Recently, I was made aware of a scientific study carried out in 1970 in the forests of Costa Rica, where a senior scientist persuaded his graduate students to be volunteers in a tadpole taste test. The scientist, Richard Wassersug, had noticed that in the anuran-diverse tropical ecosystems the ecology of tadpoles hadn't been studied in all that much depth and he had noticed that some were very dark and obvious, a bit like our slugs, others were almost transparent, and others were strikingly coloured. The question he set out to answer was, did they vary in palatability? And if so, did the distasteful nature come from the skin, the tail or the body? So in a standardised procedure the students were given fresh tadpoles to eat – the first test was to hold them in the mouth, the second was to bite the tail and the third was the full chew. The testers (who must have wanted that beer bribe pretty bad) were then asked to rate each of the experiences from 'tastes good' to 'highly disagreeable'. The results were that they did vary quite a bit, as did the responses of the students, but the universally least tasty were the tadpoles of the cane toad, a species that is of the same family as the one I tasted in my pond. They were described by the experimenters as very bitter and it seems that the source of this unpleasant sensation was in the tadpole's skin – in other words, the taste didn't get any worse once the animal was chewed. The boldness of these highly noticeable tadpoles was their warning. While the procedures here may be enough to make you cringe, the actual knowledge gained goes a long way to explain why animals such as otters, mink and herons will go out of their way to remove the skin of adult toads and why the tadpoles seem to be left well alone by many other creatures they share their environment with; it's all down to the bufotoxins giving them a highly disagreeable taste.

The same investigative technique can be deployed to answer all manner of other conundrums, such as why ladybirds are brightly coloured. Many sources will tell you that their bold coloration is of the aposematic kind, a billboard of warning, a conspicuous advertisement to all, and that what lies inside the body of the beetle will take the edge off the meal by being unpleasant to eat or poisonous. The thirty-eight different chemical compounds that lady-birds exude are unpleasant to taste. But what do they taste like, how bad can they be? Driven by that same curiosity and the need for some first-hand experience of the world, I set out to investigate. A smart individual would perhaps have started with a small dab of the yellow exudate that oozes from between a ladybird's leg joints when it is alarmed, a form of reflex bleeding, whereby the toxic blood of the insect is put out there as a warning.

Ladybirds don't like being even mildly threatened – the astringent, soapy taste is not something I'll forget and that, of course, is the whole point of the survival strategy. The badge that they wear, the bold colour and patterning that catches the eye, does so not for our own delectation but because it is part of a brutal survival strategy, hatched up over millions of years of the beetle's evolution. The ladybird is brilliant in every sense of the word.

Now, every time I see a ladybird, whether it's settled on a rosebud, tottering along the window sill or even displayed on the spine of a book cover, I can recollect that moment. The taste test, unorthodox as it may seem, has enriched my future experiences. Whether it is a dab of the blood that has just come jetting out of the eye of a horned lizard, an extravert slug, the 'blood' of the bloody-nosed beetle, or terrorising tadpoles, I have, by exposing my taste buds to it, developed a much greater insight into my world, a taste for life and its many processes.

While we're on the subject of a taste for life, our sensory testing of fungi also serves to prove a valuable lesson in this respect. A while back I spent some time with a mushroom collector in Switzerland. Though many of our Continental cousins throw themselves into life with a little more vigour than a Brit like me, I felt my English reserve justified when it came to my reluctance to positively identify as part of my potential breakfast a basket of mystery specimens I had collected on one of my morning foraging trips along the hedges, over pastures and through woodlands.

I was rather surprised when my fungus-familiar friend started to take small bites from them. On first appearance, this appeared to be rather foolhardy. However, seeing the look of horror on my face, he explained that he was just taking the smallest morsel and holding it on his tongue for a moment before spitting it out. Apparently, this method can be safely used to taste even the most poisonous species as long as you completely clear your mouth and rinse it out and spit afterwards.

It turns out that if you're tuned in to your taste buds, this is a perfectly practical technique for getting a quick idea of the edibility of your specimens. Many of the old field guides also recommend this and use taste as one of the distinct identification features of a raw mushroom (along with smell, as briefly covered in the previous chapter). The key, however, is not to go randomly chewing your way through any old mushroom or toadstool you find but to be sure you've got an edible one.

14

Getting a Feel for Things

'TAKE 'EM off, we're going on a thermal walk.' Shoes and socks were reluctantly discarded by the base of the old beech tree. As the foot attire was taken off, frowns and puzzled expressions were put on. I was leading a family nature walk and earlier one of the mums had confided to me that she thought that me walking barefoot was an irresponsible example to set on account of it being dangerous, and about the same time on the walk one of the kids had made some comment about not being allowed in the mud as he needed to keep his shoes clean for school.

I decided they were ready; this stretch of the woodland ride didn't have much in the way of holly, whose prickles, very much like the thistle's, have a habit of defeating the objective of a beginners' barefoot walk.

As the group padded along behind it didn't take them

very long before they began to respond, and within minutes they began to really feel. I could tell by the rise in excited noise that they were instantly enjoying a walk with a different dimension. Just minutes before, the younger members of the party were shuffling along, kicking their heels, scuffing the ground. If there wasn't a bird, beetle or lizard to hold their attention, they seemed bored, lacking in stimulation. Their feet, as most of us make a habit of in the Western world, were bound up in sensible boots and shoes. 'Bring stout and sturdy footwear,' it had said on the course description, and they had. But now their feet were as feet are supposed to be: out, liberated not only from the cramped confines of a shoe or boot, but also from the monotony of being a modern foot. The sensitive sole of the foot suddenly had something more than the ever-constant insole, just another bit of manmade consistency.

Now, not only were all the bones, muscles and connective infrastructure of the foot having to move around and work together, compensating for the uneven ground, the stones, the tree stumps and fallen boughs, but the skin itself was rejoicing.

Proprioceptors, thermoreceptors, mechanoreceptors and, I'll admit, the occasional pain receptor, were all firing off simultaneously, where before, little was getting through the armoured wall of rubber, leather and Cordura nylon.

Everyone was interacting differently; there were reactions to the ground, its texture, questions as to whether to avoid the mud or the gravel, how cold was the water in the puddle and with it all, an acute awareness of the environment was suddenly born. No lesson, no teachings, just a boots-off experience. The reluctance to take the shoes off at the beginning of the walk had now become a reluctance to put them back on.

This is just one of the many lessons in how we can use our biggest organ and the largest sensory interface we have at our disposal: our skin. While barefoot is best, many of us have to mix it up in the artificial world that we've constructed for ourselves, so although I'm not seriously suggesting you walk barefoot all the time, the downside of the occasional barefooted experience is that the soles of your feet don't toughen up to the point that you don't feel the thistles and thorns.

From the perspective of a naturalist, another part of modern life that creates a barrier to experience is our clothing. We constantly walk about wrapped in a glove, swaddled in fabric that stifles this particular sense. The reason we think of our fingers and hands when we talk about the sense of touch is because these are the only parts of our bodies except our faces that our modern domesticated selves, or, as I've heard us described, *Homo domesticofragilis*, have left exposed.

At this point I am reminded of the homunculus I saw on a school trip to the Natural History Museum in London. This grotesque cartoon-proportioned human model caused great hilarity and much sniggering behind cupped hands among a slightly puerile school class. The fact that it had massive feet and hands, Jagger-esque lips and swollen genitals made it look funny and the serious point it was illustrating was no doubt lost on me at that particular moment in time. This was that the sizes of the body parts, the distortion of the form of this visual representation, were in proportion to the number of sensory receptors each region of the skin typically had. After our lips, those big hands are the best-endowed parts of our bodies when it comes to sensitivity to touch. But what strikes me is our feet: they are not far behind our hands on our unbalanced homunculus. In our bipedal way of life, our feet are for the most part the only direct physical point of contact with the earth, and we go and put them in boots.

The full solution is to become a naked naturalist – or naturist naturalist, just to confuse the already perplexed. Only by exposing all of our organs of touch are we able to release our full potential as regards sense. However, let's add a dash of pragmatism to this thought: it's not going to happen, we love our clothes, our shoes, boots, fashion, far too much for this to be a serious consideration.

Many times, in many parts of the world, I've got caught up in the joy of the moment and with ill-considered bravado I've tossed aside my shoes and socks and stepped out of the mud hut with the well-meaning intention of emulating my barefoot local guides. Inevitably, while they stride on, I'm left hobbling and hopping and humiliated in their wake, stopping frequently to remove plant spines, chiggers, ticks and dislodging sharp stones from between my toes; it's frustrating. I've even deliberately shadowed these earth walkers, treading only where they tread, and I still end up being hobbled by the experience. While my local friends might stop to draw the odd serious offender – a palm spine or questing tick that's got a foothold – for the most part they're on the move.

Feet fresh out of boots are pale and soft, the skin moist and a little too vulnerable; they've also been cramped up, toes are piled on top of each other, creating an irritating niche for debris to catch in – a healthy, wide foot, unfettered by footwear, has less of an inclination to do this. However, choose your moments and you'll find you gain a lot from the experience. Stalking wildlife, moving through the environment, assuming you don't blow your cover and scream when you put your foot flat on a rosette of thistle leaves, is a dream, you can really feel the ground. Not just the twigs that might snap, although with a little practice you can tell a green springy twig from one that is dry and likely to create an explosive failure and you can sweep leaves out of your path with your toes. Walking barefoot allows you to move as quietly as if you were floating; it takes field craft to another level.

If you are new to the world of 'bare-footing', one of the sensations that you'll be aware of once the novelty of feeling mud between your toes has worn off are the thermal qualities and the variation in the temperatures of the terrain. It was and still is one of the more subtle pleasures, a sensation from which your thermoreceptors have been insulated. Feel the difference between the cool grass at the edge of the path, which effectively is a forest in miniature complete with its shade, and the mud. Shadow the water-cooled blades of grass absorbing any of the sun's insolation and hiding it, and then feel the hard, barren tracks of the cracked mud. The answers to questions you might have regarding the places that lizards bask and tiger beetles scurry and glitter are right there under your toes. I once even stumbled upon an adder acting oddly. It was coiled up on the edge of the path, not in the sun patch, no more than a few centimetres away, which would have been more typical but in the shade. It was close to a complex of roots and I assumed that it was on its way either in or out of its hiding place.

As soon as I had spotted her, I froze and then, centimetre by centimetre, I crept closer and closer. It took the best part of fifteen minutes, and by performing the sort of posture control that my Yiquan teacher would have been proud of, I found myself just centimetres from her glorious gargoyle face. This act was helped considerably, as it often is, by being barefoot. What puzzled me, and anyone who has spent time with these shy reptiles will understand this, is why she stayed. She had formed a flat coil of her body and if the sun was on her I would have made the assumption that she was basking, absorbing the solar energies for her own internal purpose. But here she was, neither in the sun nor under cover – very un-adder-like.

Eventually, and I like to think it was naturally, and not because of my presence, she became animated. A twitch of the head and some surreptitious tongue flicking, and she took her voluptuous, tessellated form off and threaded herself back into the tangles of the nearby roots. I got up, but before I moved off, a thought came wandering into my head, an unformed question, stimulated by what I was feeling in my feet. Rather than bend down and use my hands, I slid my foot to where she had been. The question was answered. My adder had been sitting on a hot spot created earlier when the sun had been bearing down on this section of the path. The ground was still warm, despite being in the shade when I stumbled into the scene. In the adder's world, why track the sun and leave the safety of your bolt-hole, if you can take second-hand from the earth what the sun left behind: it was a mystery solved with my feet.

We might think that by putting shoes on our feet we're taking a cultural step up the progress ladder. Underneath it all our feet are very well endowed, something we have left over from our tree-living primate ancestors.

Imagine a normal working day, cutting the bread, making a cup of tea, driving to work, typing or operating a machine, but the only difference being that your hands are in socks – not only would it be difficult to manipulate anything but you would also be missing out on many sensations that you simply take for granted every day. This is, effectively, what you are doing with your feet by putting them in shoes all day.

Most exercises that explore our sense of touch involve our hands. These glabrous and dexterous limbs are very much our go-to feeling organs and a lot of fun we have with them too. I get a lot of simple sensory pleasure touching things, we all do. If we're close enough, say within an arm's reach of something, it's one of the first compulsions any human has – to reach out and touch.

The moment I catch a snake to show to a group, present a small mammal from a live trap, or remove a skull or a shell from a box or pocket – whether the group is composed of school pupils, teachers, families or grey-haired adults with a lifetime of sensory experiences behind them – there will always be someone who reaches out with questing fingers. It's natural. We live in a world that is forever telling us not to touch, yet it's such an important faculty at our disposal.

Can you identify a species of tree by touch? Trees are one of the most abundant, accessible forms of natural textures and they show a lot of individual species-specific variance, enough to make a great exercise out of exploring your somatic potential.

A fun, demonstrative game to play with a group of kids or adults is to give them a few moments to find a tree they like the feel of and leave them to their own explorations, drawing attention to the way the bark is textured. The topography of the bark is related to how specific trees grow and twist; get them to feel the scars, the scales, the lenticels: the micro and the macro landscapes. I imagine I'm the size of a bark louse, an adventurer, a 'Cortez of the cortex', and I'm trekking across the bark's mountains and valleys. I then get the group to identify the tree and then, one by one, I take the blindfolded individual to a number of trees and get them to repeat the journey but on several different species of tree, a few minutes each. They have to identify, as soon as they are sure, the type of tree that they first chose. It's quite revealing how good we are at it. When we focus on our somatic senses we are super-sensitive, able to decipher the Braille that life surrounds us with, right down to details of 0.2 millimetres, if we simply touch, but if we track – that is, slide our fingers, something called dynamic touch – we can improve on that resolution and perceive texture irregularities on the surface as small as 13 nanometres or the size of a large molecule! While not quite as sensitive as our lips, our fingertips still pack in over 2,500 touch receptors for every square centimetre, and our feet are not far behind.

The world of feeling and touching is so vast. Everything we've talked about so far is wrapped up in this one sense; it is omnipresent and all-encapsulating. Your skin is your biggest interface with the world around you. It is loaded with millions of receptors that relay information to your brain all the time. Its influence on everything we do is so vast that pinning it down to specifics is beyond the realistic realm of what we are trying to achieve here. But, needless to say, those five million or more touch receptors are pretty useful tools, whether it's feeling the tiny, gripping jaws of nematocysts as you pull your finger through the tentacles of a sea anemone, feeling the sun on your back, detecting the thermoclines when you dive down to the sea floor, testing a branch, feeling a tree, or stroking a bird's wing. Those two square metres of skin that we all possess as adults are never to be ignored or underestimated. Like so many of the senses, if we are consciously aware of our natural abilities in this area it can only help us to derive more from our explorations.

15

The New Attuned You

You're a proud owner of the most important toolkit you could possibly ever own, and it's the one you were born with. Assuming you're new to all this it might feel as if you've just got the tools out of the box. You might have tinkered with them a bit and checked that all the bits work, and now comes the moment of truth, the fun, rewarding bit, the whole point of what this has been about – it's time to take yourself out for a spin and use your new-found awareness to its full effect.

Put all the ideas and thoughts presented in the previous chapters together and apply them, and, by definition, yourself, to the wild world. This thrill is free and it costs you nothing but time, but it has the potential to give you back bucketfuls of benefits.

It's good not only for your mental health, but also for

your physical health, it gives you insight into an ecocentric, more natural way of existing, and by increasing your own awareness of the natural world, wherever you may find or experience it, it opens your eyes up to the influences that we as a species have. When armed with this knowledge we have a good basis for making decisions and minimising our impact and for improving our future relationship with the world of which we are a part.

Seeing the beast

The bird just vanished. One moment it was perched on the rock, a king on its mossy castle isolating it from its riparian backdrop, a moat of wild, frothing, kinetic river water. The next, it had spread its wings, but rather than purr off in the bird-like manner that would seem appropriate, this bird, a dipper, dropped headfirst into the raging riffle. Within the blink of an eye it had gone, had vanished beneath the surface.

At the time, my young mind was having trouble computing all this. Small birds just didn't do this sort of thing in my picture books. Ideas no doubt buzzed around my head. Had this been some awful accident? Had I witnessed a strange suicide? Was the bird sick or had it fainted?

Then, after what must have seemed an eternity, up it bobbed again, like a cork held deep in the bathtub. With a blink of its nictitating membranes (a very special third eye-lid, which shows up as a white flash to the distant observer, but acts as diving goggles to protect the eyes when it is foraging) it was back up on the rock. A quick flick, sending a shower of mercurial water droplets tumbling, and it assumed the pose, almost tempting my disbelieving mind to think that I had imagined the whole incredible thing. This was probably my very first birding or, for that matter, wildlife-watching experience. I would have been maybe three or four years old and I had accompanied my dad on a fishing trip in Braemar, Scotland.

I remember the moment as clearly as if it was yesterday; it was a spellbinding, seminal moment in my life: a wild creature, a mere fishing rod's length away from where I sat on the river bank. Its trusting presence permanently altered a part of my brain. Somewhere deep in my hippocampus things would never, ever be the same again.

Frequently, I find my life as a modern human dull, burdened by responsibility and the drudgery of paying bills and taxes, and, like many of us, a lot of my time is concerned with mundane decisions. I find myself obsessed with details that really, in the greater scheme of things, are not that important. I get a bit depressed by the state of the world, the politics I don't really understand, the fact that nobody but me, it seems, is concerned about the felling of that tree, the abuse of that hedgerow, the amount of plastic I am forced to discard. These are fundamentals – not the fripperies that society's lack of consideration makes them out to be.

It's the cost of being ecologically aware – things get to me, they niggle and needle into my happiness and, ultimately, get me down. As I sit, tied by necessity to a computer and keyboard as I often am, relentlessly hammering on the keypad, one finger at a time, putting down words in some kind of an order, more often than not I'm describing the things I want to be outside looking at and experiencing. I'm not alone in this, I know: we would all rather be on the beach, sand between our toes, feeling the dry rattle and buzz of a heath, or slumped on a velveteen mossy rock in an ancient oak woodland staring at the canopy. But, like me, you're probably stuck in the humdrum of being a modern human, in an overly comfy cocoon, numb to experiences that truly stimulate the senses we've talked about.

A lot of my material, I'll be honest, I collected when I was ten years old. It's an irony that I exist in a world where the younger generations need to buy a book about, or go to a forest school to learn about, things that were, only some fifty years ago, free and part of a common knowledge. I write about things a nipper in the 1950s would have taken as a given, and it was free for the taking too. These children didn't need someone to show them the pleasures of finding frog or newt spawn, birds' nests, big caterpillars. These kids would know which trees made the best spears, whistles and firewood and which were the best for climbing or which gave free snacks. We're losing those experiences, that natural knowledge which we have as a species.

I often find myself getting into a downward spiral and right old funk about my species and its predicament. I want my daughter to know a better place than this, then I get maudlin about whether anyone will read this, does it make any kind of a real difference, and then, just when I've reached a new level of low – a robin lands on the feeder tray that is stuck to my window.

It flicks its wings, its red breast an ember to my tinder-dry soul, it cocks its head and looks at me with those deep dark eyes, and that's all it takes: I'm alive. It's my bear moment again but this time in miniature. As I look, I'm seeing, and as I see, I fall into its wild world. This moment threads together every moment I've ever spent with a wild creature. It's part of an interconnected whole, a life experience, and it has come full circle. It started like this nearly forty years ago with my dipper and it's taken me all around the world. I've similarly stared into the eyes of shark, whale, gorilla, cobra and hornbill and here we are again. All of these experiences are very different but at the same time they share a common fundamental: they all move me and pull me in closer. This robin is not my robin, he just shares the space with me, living just the other side of the glass. The robin makes my soul soar, it lifts me in some inexplicable way, makes everything right. This simple contact with a wildling puts all the other nonsense in life into perspective. All the things that worry me in my small world get thrown into the part of my brain's mental filing system that is labelled 'superficial'.

This robin – but it could just as easily be a beech tree, or a woodlouse – and whatever sensory door I choose to let him in through is something raw and powerful and encapsulates what it is I'm trying to get across in the pages of this book. If I could just find the words to describe what happens in this moment when wild comes knocking, it would save an awful lot of typing.

Mr Robin is so close. I trail my eyes over the different textures of feathers, each with a specific job to do, and I notice the strange arrhythmic pulsing of his chest and wings. Then he bobs and grabs a mealworm and makes a very quiet call, and is joined by another, its mate. I now know it's 'Mr' Robin as the second bird starts to play all soppy and begs like a chick. He feeds her, stuffing a mealworm in her gullet. This is courtship feeding, they're a pair and he is feeding her up, making sure she's full and able to give everything to the gruelling process of laying eggs which lies ahead.

Nature just let me in on one of its secrets, but more than that, there's a connection here to a long line of robins – the ancestors of which messed around opportunistically grabbing insects disturbed by the hooves of wild boar and the aurochs in the great post-glacial wildwood. Have no doubt, this bird is a little robin-shaped shard of wild and we're in and amongst it. If you just open up to it a little, you'll see it everywhere.

Any contact with nature like this can provide the sort of genuine, bona fide thrill which is becoming increasingly rare in our modern world. Not rare in the regular sense, the raw materials are all around us, but rare in our pure and simple appreciation of it.

To see a wild animal – I mean to really see one, to watch it go about its business, to experience it in all its wildness and become a part of its world and not a cause for alarm, the threat, the thing that sends it scuttling off into the safety of the great beyond – is one of life's *real* experiences.

It's a way of pitting your wits against the wild and it's not really something that happens by accident. What I mean by this is that you have to put in the groundwork, you have to be open to the experience, and in order to do this you need to hone your own skills to a level that might have been familiar to our ancestors. If you like, you are finding your inner functional hunter-gatherer and you are then deploying him or her to the pleasurable task of getting as close to your quarry as you possibly can.

It's these intimate moments with the natural world that last; seconds spent watching a diaphanous butterfly unfold from its chitinous prison, or a moment in the presence of a badger as it goes about its business in complete nescience, are experiences that you will hold on to and replay over and over again. In my experience, it is these real moments that outlast and endure, not that bargain you got on eBay or the highest score on Donkey Kong 3. There is a school of thought that says I'm wrong and that it is only experiences that are important to us that change the synapse patterns in our brains and cement themselves as a permanent memory. So it could be argued that if you dedicate all your life to achieving that highest score on a computer game, then this will become your top-trumping memory, simply because it is important to you. My argument is that a wild experience in nature, if we allow it to happen, lights up the neural pathways that lie dormant or underdeveloped; we have this natural value system deep inside all of us and a genuine moment actually engages with something deep in us, so much so that after a magic moment we feel a real sense of purpose and deep satisfaction.

Our artificial, safe, increasingly urban lifestyle is a pretence of a world, manufactured to our own specification, making the processes of our everyday life as easy, clean and comfortable as they can be. There is little in it that represents a real taxing challenge, other than the sheer mental will to carry on doing the same thing day in, day out. There are no exercises for the innate beast in us all. To find this real significance, to discover the point in it all, we need to explore and find the edge of our comfort zone, to go out and make genuine connections with the world. It's about reconnecting with our own inner beast, the creature that lives within us all, by immersing ourselves in a world where the wild things are.

16

You Can Never Rush a Snail

WHERE DO you start? Assuming you have no experience of watching wildlife, this whole subject matter of getting close to, and reconnecting yourself to, the wild may seem to be a bit daunting at first. For a start, you've got millions of species of other animated life forms to choose from, and you'll almost certainly have in your imagination the fantasy image of what you might think this book is about. But trust me, there is a point to this – it's the beginning of a journey and it's a journey that starts off at a walking pace.

While you might hanker after being familiar with a fox or matey with a moose, or even gaining the confidence of a Kodiak bear (and any other wildlife alliterations of which there are many), the reality is, it's probably better, more achievable, cheaper and indeed safer to train your senses with smaller and more accessible life forms. The little things in life rarely let you down and few will eat you if you get things a little wrong.

Before tigers, black rhino, badgers, dippers and even robins, it was caterpillars. There is seemingly little in common between a top feline apex predator and an unassuming fleshy grub of a thing, but there is. This personal connection with nature for me started in a jam jar. I have to reiterate, in a world of flow diagrams and step-by-step hand-holding instruction manuals, you don't have to emulate my path; this is merely a serving suggestion. Find your own thing, but by all means, if you're struggling for inspiration, you can do a lot worse than picking a few large white caterpillars off your nasturtiums or cabbages and observing them.

To take a small creature and look at it through a different set of eyes, to smell it, touch it, feel it, is great training – you can do all these things, surprisingly, with these caterpillars. For a short moment in my life, all of my focus was on these 'pets'. I did all of these things, I drank them in in every way I could (almost: I never tasted one).

You might not think of a caterpillar in a jam jar as a multisensory thing, but I just have to see these mottled eating machines on my allotment and I can recall watching them moult (five times), the periodic struggle to rush out of the previous tight skin and reveal a new one underneath, the smell of the 'mustard oils' as I unscrewed the lid to change their food every day, and the same smell from the green goo they would vomit onto my fingers if I was a little rough with them, I can even hear them eating, a methodical, miniature, mechanistic ticking. I had always thought the caterpillar built its chrysalis around itself, but I watched an animated caterpillar wriggle out of its skin to reveal something unlike its previous form. The mask was removed to reveal a bag with no recognisable caterpillar features, then it hardened into a box, one that wriggled when gently touched.

Finally, the climax of the whole life cycle, the crescendo of wing and limb, came when, a couple of weeks later, that everyday hedgerow miracle occurred – the chrysalis ruptured into butterfly life.

If that's not enough to stir you up, to bewilder and amaze you and to generate a list of questions, then I don't know what is. It had me hooked and then I started to look at butterflies, not just as floaty comments of summer, but as marvels, each one an envoy for a magical metamorphosis, four different life stages, each with its own mysterious transformation, each with its own ecology, predations and dramas.

They got me out, I was a caterpillar hunter. I poured over details, honed search images to hunt out their often highly cryptic forms; I stalked the adults as they nectared on flowers, just because I could. It was a game at first, but then I was mesmerised by their patterned eyes, the mechanics of the watch-spring proboscis. They drew me in and while I fell into their rabbit holes of wonder, I found other things, denizens of the world, which equally challenged the senses and my interpretations of the world at large.

When was the last time you got down on your hands and knees and simply stared at the world of the grass-roots jungle? I mean really got eyeball to compound eye with a beetle or an ant? Most of us go about our lives completely unaware that we are missing all the good stuff that is going on almost under our feet. In fact, when we notice these most numerous groups of creatures on earth, underfoot is pretty much where many of us would want them to be. I've never really understood all the negative connotations which are conjured up just by the mention of their names: fly, spider, caterpillar, moth – you get the idea.

However, one of the real challenges of life is being able to see these inhabitants of the microcosm as equal denizens of this planet; every species, if we're talking about the concept of biodiversity, scores the same.

By this I mean that if we're talking about the importance of the variety of life on Earth, that big fluffy giant panda is as important as that ant that just scuttled across the paving slab and disappeared down the crack. The only real difference is our attitude, knowledge and understanding and, sadly, it's harder work to try and relate to an ant, and it's a whole lot easier simply to hate them.

Ant or anteater, tiger or tiger moth, every living thing (and yes, that includes wasps, mosquitos, slugs and spiders) has a function within its ecosystem, and part of a naturalist's ability to connect to the world is to not only understand this interconnectivity of life but also to go searching for some of the answers. Admittedly, it can be a bit of a challenge, but with some sound observations and a curious mind, you can become acquainted with some of the most fascinating of animals and, in doing so, connect at a much more intimate level with the world around you.

The best thing about the invertebrates is that small creatures take up little space and so there are always some to be seen wherever you are and at all times of the year.

One of my favourite activities, one which really helps me to get my eye in, is to do something that in our hyper-fast modern world there never seems to be much time to do, and that is to lie still. Find some nice, tall grass and, after a quick check for thistles, nettles and other objects that it would not be desirable to lie on, get down on your belly, get comfortable and simply watch.

Just stare at the grass and other plants and, deliberately, slowly pull focus through the depths of this verdant miniature forest. Sometimes it helps to partly squint through your narrowed eyelids, as this cuts out peripheral distractions and allows you to pick up on any movement.

Recently I did this with a group of retired professionals who, when they came on one of my wildlife-watching weekends, mistakenly believed that the sort of wildlife watching we were going to do was all from the comfort of a minibus or at least while standing upright. However, after a few minutes of giggling and grumbling as arthritic limbs were bent and buckled into load-bearing contortions – some shapes which, probably, hadn't been experienced by their bodies for some fifty years – they settled down. Where moments before there had been a group of eighteen or so octogenarians, bedecked in red and yellow cagoules, there was suddenly nothing visible or moving. Save for the gentle breeze rippling the grass and causing the wild flowers to bob around as if freed from their earth-bound roots, it was for a few moments at least as if they had been suddenly atomised and beamed off to another planet. Well, they sort of had. Slowly, as old eyes became accustomed to the rhythms and oscillations of the plant stems, they were transported to a world inhabited by multi-legged aliens and creatures.

Strange cooing noises, gasps and the odd slightly more abrupt exclamation signified to me that somewhere in the sward in front of me my group were witnessing things never before seen.

As you stare into the tangle of greens, the exercise starts with colour. Immerse yourselves in the multitude of hues and tones of this rarely witnessed world. Really get tuned in to the variety of pigments that are on display. I like to think about what it is that makes these colours all so different, the thickness of the leaf or stem, the other microscopic structures within that either bounce light around, let it through or obscure it. I imagine all those science lessons about transpiration, photosynthesis and circulation in plants coming to life. This is what fuels life on Earth. Inside each plant, through tiny pipes and micro-plumbing, the elixir of life is flowing, magical sugars and proteins made from nothing but light, water and the gasses we all breathe in and out. Conjuring up an ecological vision of the molecular alchemy in your mind's eye helps with understanding what the others are up to.

No sooner do you start to imagine the building blocks, the cellular machinery and the ebb and flow of microscopic living processes than you meet one of the giants, plumbed in to a stem by its sharp, piercing rostrum mouthparts; a plant hopper feeds like a wine thief stealing its sugary needs.

You hadn't noticed it when you first pushed your face into the grass, but now, as you've tuned in to the minute, rhythm and patterns of the plants, it suddenly stands out; its mere shape and form, even though beautifully matching the colours and lines of its environment, appears to pop out at you. Just the swell of its body, the twitch of a leg or antennae is enough to give it away to your now-rejoicing senses – you've just taken a first step into a deeper natural awareness.

In no time at all you'll be feasting your eyes on a world you had probably never seen before and one that you may not have even experienced unless you had undertaken this, the simplest of instructions – to take a little time out and lie down on your belly.

For some of us the planet feels as if it holds few surprises; the great explorers of the nineteenth century and the modern scrutineers of science sometimes seem to have stolen all the real exploration and excitement for themselves. Sometimes it's easy to feel that we might have been born 150 years too late. Not true. The perception is a false one – to become an Indiana Jones, an Isaac Newton or an Alfred Russel Wallace you need do nothing more than change your perspective on the world, lower your horizons and get down on your elbows.

It's as easy as that, you just have to build that bridge in some way. I found that insects are very accessible and you can fast-track into the wild in their company. But, of course, they are just a small part of the equation. They don't exist in isolation, nothing does.

Your effectiveness improves with every curious moment. If you've taken away some of the techniques in the previous chapters, you will inevitably observe, smell and hear other things. Question and quiz all that you experience and you'll start a magical mystery tour.

My thing for caterpillars led to a basic knowledge of botany – it had to, you need to know what they eat, in order to find them. Some are not that fussy, while others are so specific that it's the limits of the food plant, not the insect, that make it rare.

Very quickly you develop an ability to identify different kinds of plants by their foliage, flower, bark and growth form, you can't help yourself. It's botanising. Not that I would call myself a plant expert. In fact, I'm not an expert in anything; 'expert' is a dangerous word that suggests it's a final achievable goal. When you hear it used to describe someone, all it says to me is that they have pursued one particular area more than another.

For the curious, nothing exists in isolation. Caterpillars, by necessity, lead to plants, and adult butterflies and moths need other kinds of plants to feed upon, so you get to know these too. Then there is the morphological circus that every stage in the life cycle takes part in – every scale on the wing, every spine, flap or flange has a role to play in an insect's survival; this then leads you into the world of what threats to survival there are, what predators – the birds, the shrews and mice, to name but a few – and so it goes on, connection after ecological sensory connection. You become ensnared in a net of wonder and with it comes an awareness – what I refer to as rewilding connection. You are using everything you've got on the world – the things you see, smell, taste and feel – an alchemy of senses, and your brain and you become part of it. The deep, core satisfaction that comes with it is the real, wild you.

The sphere of influence

Something to bear in mind, something that neatly encapsulates all of what we've learned about ourselves and the world we're trying to immerse ourselves in, is that although we may not for the most part be a threat, we move clumsily and noisily about our business in a short-term 'there's nothing around that might eat us' kind of way, and are unbelievably disruptive and disturbing to all around. As a consequence we spoil many a moment before it ever happens.

Part of my work as a naturalist, whether I'm working on a film and waiting for the shot, doing ecological survey work, or simply enjoying nature in a meditative way, requires me to spend quite a lot of time, maybe more than most, simply sitting still, being patient and waiting for life to happen all around me. It's the art of sitting still and being able to shut up. There's nothing to it, it's not rocket science, and yet it's something we rarely do. Try it, find an hour a week – it could be your lunch break, or a detour on the way home from work. Just find a spot, and sit there. Practise tuning in, listening, looking, smelling, all of which can be done while rooted to the spot.

On several occasions I have been sitting quietly and have witnessed first-hand other people 'missing' some pretty amazing stuff simply because they are crashing through life, turned off to possibility. I'm always hearing how people are amazed at the luck I have, the things I see on a regular basis, yet there is no reason why they can't experience these things too. While luck always plays its hand in whether a moment is experienced or not, it cannot apply to every scenario. I put a lot of this down to something I call our personal sphere of influence.

When we go for a stroll, it's a fact that we have a rolling, invisible effect on all that we pass. A sphere of influence on all the life forms that surround us, especially those with a well-developed central nervous system. If you walk quietly, calmly and slowly through the landscape, your sphere of influence shrinks. Conversely, if you speed up, your distur-bance factor increases proportionally. When you are moving slowly, you will also tend to notice the details and the finer nuances of the world around you, and this is when you get to see stuff happening. Stop moving altogether and let a little time go by and your invisible footprint can sometimes almost completely dissolve; this is when butterflies land on you, mice run over your feet, and birds perch and sing, so close that you can hear their breath.

This increased ability to notice things is partly down to your lower speed: you've just given your senses more time to take it all in but you also represent less of a threat, you make less noise, and are able to adjust to the unfolding picture around you: a stoat moving its kits makes you pause, enabling you to drink in the furry wonder of it all; a pied flycatcher snatching caterpillars from bursting oak buds – you stop, it stops, and after a breathless moment of mutual scrutiny, its assessment of your threat level enables it to continue feeding. Such trust gives you an insight into an intimate everyday process.

When you started this book, you might have been one of many people I meet who are desperate to connect with nature in some way but have been put off somewhat by the lingo, the bewildering array of subject matter and, perhaps, the consequences of getting something wrong. Expert syndrome also takes its toll on some people, clipping their enthusiasm, drowning the sometimes childlike wonder in a sea of seriousness. Hopefully, many of these fears have now been allayed. Boil everything down to its basics and take each moment for what it gives you; drink in, absorb the sensations and revel in the experiences you have on your adventures. You may well find yourself lost in a taxonomic maze, unable to put a name to a face or feature. No matter. We often get ourselves in a tizzy over the identification of something, a plant, a bird, an infeasibly small moth – I know I do on a regular basis. If I've been on a fungal foray or have just emptied a moth trap and am working my way through all of my specimens, I always have a handful that make me want to cry, give up and take up a simple hobby like stamp collecting instead. Take comfort in the fact that every 'expert' has times when he or she feels this way. In a moment of insane frustration brought on by a field guide to fungi, I tore it up and fed it and the little brown mushrooms to my giant African lands snails – they, at least, got some sugars and starches from the experience (it wasn't a very good book, as it turns out). Paying attention to the details and recording what you experience will take you a lot further. This familiarity is what brings the answers in the end. Eventually, some of these skills will become first nature and in time you'll start to unravel the details.

Foraging is a good example of this – wild foods are everywhere. Finding them, identifying them and eating them is a very satisfying way to connect with the natural world. As ever, how things unfold always boils down to being careful, taking your time, using your senses, exploring and trusting what you're seeing, smelling, feeling and tasting. The basic skills and lessons in this book should help empower you in whatever natural adventures you wish to set out on, whether it's trying to get that perfect photograph, meal or insight into a wild life.

17

Why We All Need the Wild

'Brock of the clan wild'

THE VALUE of this connection with nature is apparent to all that make the jump, and even to many that don't think they have anything to do with the wild, especially when they get down to thinking about it. That moment when you kick your shoes off in the park on your lunch break, the uplifting song of an unnamed and often, I suspect, only subconsciously noticed bird that puts a spring in your step, and we all crave a holiday, don't we, a weekend by the sea or a walk in a cool woodland on a summer's day . . . all these are what we are meant to feel. Yet when I look around the world, I can clearly see symptoms of what some ecologists refer to as the sixth extinction, an anthropogenically driven loss of species and an unravelling of ecosystems – something that we humans are entirely responsible for. We live in a time of daft political decisions, the call for culls of otters, gulls and badgers. Ecological mitigations to alleviate the guilt of another 'much-needed' housing development, or some other false progress, all at the expense of something that is priceless and that we need in so many more vital ways, not just the 'ecosystem services' that they provide.

The anticipation was killing me, and the midges and mosquitoes that had plagued us since the sun submerged into a pool of tangerine light were doing everything in their power to do the same. Exsanguination by a million bites.

For an hour and a half, I had perched on this old fishing stool, desperately trying not to fidget and cause its tired aluminium legs to creak. I had fought numerous sneezes; I had gently blown away, through tight lips, insects from my face; I had staunchly resisted the biting midges burrowing into my inner ear; staunched several burps, although the borborygmi had got the better of me – there's not a lot anyone can do about burbling stomachs. My grandpa was the other member of this intrepid badger-watching duo – he had foolishly agreed to accompany me into the twilight.

So here we sat, watching and waiting, under an old oak, looking down the embankment of a disused railway; each willing the other to crack first, call it a night, and then we could head home for hot chocolate.

This wasn't quite the image in my copy of Gerald and

Lee Durrell's *Amateur Naturalist*, the simple line drawing of a boy in dark clothes, back to a tree with badgers foraging a few metres from his feet. It didn't quite tally: for a start, it couldn't convey the agony of blood pooling in buttock cheeks, the creeping cold working its way along his bones to his core, and it certainly didn't illustrate the clouds of bloodthirsty insects that would have been cavorting around his head.

Then we both heard it: a scratching sound. Impossible to describe, but if you rapidly scrape the hairs on the back of your neck with your fingernails, you get a pretty good simile of the noise emanating from the tangled darkness. Then again, six, seven times in each burst. I strained into the darkness, suddenly piqued. My heart was racing, my mouth dry as I desperately tried to decipher shapes in the dingy understorey light. Then there it was; I saw the white stripes of a humbug head, bobbing in and out of sight before slinking away backwards, dissolving, absorbed into the inky night. We never saw any more badgers that night and it really didn't matter. It was the most exciting thing I had ever done.

I was about eight and from then on badgers were my everything, they were my portal into the wilds of the woods. They were my interface through which to relate to the world. While I was watching badgers or, more often, waiting for them, I became a part of the woods. I saw and heard everything, like they did. They were my teachers, my stripey, emblazoned tutors of the wild. I learned through them, not just about birds and their song, but I got to know their rhythms, their other lesser-known calls and vocalisations;

mice ate peanuts from my hand, rabbits got closer than ever, fox cubs played just feet from my feet, a mole even ran across my legs.

When I was in their presence in the hushed solitude of a wood at dusk, I was whoever I wanted to be, I was a tracker, I was Grey Owl, Grizzly Adams, Gerald Durrell, Alfred Russel Wallace, I was a time traveller. They seemed as ancient as the hills themselves. A raw end to a continuous unbroken link to the ancient wild woods, I imagined them here, in this same precise spot, sharing their world with other larger mammals long since passed.

It was a real adventure, one I could have any day and one that made me me. I would try and get out to see them at every opportunity. These mammals and my almost daily moments with them gave me the same joy and anticipation as seeing my friends at the school bus stop; they were part of the very fabric of my younger life.

At first I tried sitting on a stool, the creaky aluminium one, then on the ground, where I felt too vulnerable and where the cold crept at me from all directions, then, after having my cover blown by changing wind direction, I decided I would wedge myself in the crotch of a tree bough, which was fine to climb up into an hour before sunset, but was a dreaded leap into the dark once it was night and I needed to get home. Eventually, I constructed a high seat out of an old wooden ladder and a chair seat and back; I then had a comfortable and portable option – if the badgers decided to move their activities to somewhere else I could follow. The intimacy I described in an earlier chapter came later; once I had gathered experiences, learned to work the wind direction and gained in confidence, I came down to the ground again when I wanted – only falling back on the high seat when getting to know a new clan, if I had been absent, or if I wanted to make observations without my own presence being a distraction.

Over the years, I got to know the clan pretty well, I gave them names, watched the cubs grow up, fed them peanuts, even became an honorary badger. When a badger wipes its yellowed bottom on your jeans, it might not be apparent, but the pungent, greasy stain left behind is the biggest compliment a badger can give you; it was like being given the key to their world, an olfactory badge, that said I was trusted, I was one of them, I was a brock of their clan.

Looking back, I now realize these moments were my time to process thoughts, get away from the claustrophobic stifling indoors, away from parents and school work, away from judgements and difficulties. These animals and their environment had become a kind of back-up surrogate family, their fields, woods and ditches my second home. I really did know the landscape in the way others might know their house, the furniture layout or the planting in their herbaceous border.

24 November 1987

When I was a teenager, several years after I had really discovered and got to know my badgers, I experienced an emotional trauma that turned my world on its head.

I remember that harrowing evening like it was yesterday. The evening of 24 November 1987 was a cold and wet one. Ready for bed, pyjamas and dressing gown on, I was watching TV in the house on my own. Dad was at a cycling club AGM and my mum had gone out a while back to pick my brother up from his weekly football practice. I remember thinking that they should have been back ages ago; it was past my bedtime, past my brother's bedtime. Where were they? What was taking them so long? Of course I took advantage of the opportunity to watch even more TV than I was usually allowed. Guiltily, I would keep taking quick peeks through a crack in the curtains to check for car headlamps while keeping half an ear open for car noises on the drive. It eventually came but the engine tone was different, the headlamp pattern wasn't that of my mum's car. Cautiously, I went to the door to look out; I'd been warned about strangers but instead of the familiar white panelling of the family Mini there was orange and blue, the unmistakable livery of a police car, reflective strips, loud and silver, reflecting the outside security lamp light back at me.

There was an apologetic and solemn look in the officer's eye that told me all I needed to know. Something terrible had happened and because they had asked me to phone my dad, I knew that the something had happened to my mum or brother or both. We waited, I boiled a kettle, enough for three distracting, awkward time-filling brews. That was the longest cup of tea I had never drunk. Dad finally arrived home with a heavy sense of dread and distress in his eyes and in the handful of minutes that followed before we were off to the hospital I learned that half of my family were more than half dead – my mum was in a critical condition and on life-support and my brother, who had sustained serious head injuries, was in a coma.

What followed was a lonely period, weeks dangling over a chasm of misery, a time of not knowing: not knowing if I would ever have a complete family again, not knowing if my brother would wake up, and if he did, whether he was going to be the 'talking head' the neurologist had warned us was likely. My dad was always away at the hospitals, the house was always cold and empty when I got back from school; there were food parcels from caring neighbours, next to neglected milk bottles on the doorstep. I heard my dad, who always struggled with emotion, crying in the night, and caught him red-eyed on more than one occasion; that former pillar of my life was unable to convincingly tell me that everything was going to be all right. The only thing I was sure of was that I was, completely and utterly, on my own.

My happy, middle-class, perfect little family idyll had, in a split second of inattentiveness by another driver late that wet autumn night, been turned totally upside down. It was so painful that I remember wishing it all to be over; I wanted them to die and release me and Dad from the pain of it all.

I tell you this because even though I lived in a caring community, had family close by and had very supportive teachers at school, what I learned in that cruel life lesson was that nobody can really make things better, they can hug you and bring you food, and talk, which, of course, undoubtedly is a big part of dealing with emotional trauma and is essential to the development of coping processes, but when it comes down to it, you've got to do the healing and the surviving yourself. Face-to-face contact with anyone was always a little awkward, everyone tiptoed around the issue, never quite knowing what to say; there was always a huge elephant in the room.

I share all this because, retrospectively, this period of intense emotional upheaval and pain taught me something about our human spiritual connection with nature that in my experience is completely undervalued by our Western society. Nature heals, it has restorative powers. The woods and fields and all they contained, especially my badgers, were the constant unmoving point of reference through it all. It was the clan of monochromatic friends in the night that gave me focus, gave me the space to process the enormously complex emotional baggage that life had just rudely dumped onto me. Thinking back to those days now, I've realised those badgers saved me hours of therapy, maybe my life? They were my reconcilers, my non-judgmental council and silent stalwarts. They, and the woods and hedgerows, the fields and ditches that contained them, were a constant, an immutable part of my life. When I would slam the kitchen door and 'run away' from it all, I was doing the exact opposite. I was running to the one thing that made sense to me, the one thing that didn't tire of me, didn't mind if I shouted at it, or cried with it.

In short, nature was good for my mental health. I still rely on my nature time in the same way. Whether it's a row with my wife, an unpayable credit-card bill, death or stress of any kind, I take to the hills, woods, fields, hedgerows, parks – wherever I can find nature I find my salvation.

I often ponder this particular value. I wonder about how many hours of therapy I've saved myself, how much time and money. If I hadn't found the wild, where would I be? How would I have coped? Could I have coped?

I've since met others who have found, by accident, the same sort of comfort in the non-judgemental natural world. I've also become aware of many who are helping people make this connection for just this purpose.

18

The Art of Rewilding

IT'S LONG been appreciated that simple contact with an animal brings with it many benefits.

An animal that is relaxed in our presence makes us feel that all is well in the world; the fact that a fellow creature is content and not alarmed and is non-judgemental puts us at rest too – this original context of 'happy' could well hark back to an ingrained survival mechanism, similar to the sound of birdsong in contrast to alarm calls. This is possibly part of the reason pets make us feel good.

Animal-assisted therapy – in this case, using domesticated or captive species – is well known and long practised. Early records of the benefits of close animal contact go back to the late 1700s with the Quaker Society of Friends Retreat in York and the rather better-known Florence Nightingale; both recognised the value of animals in treating the infirm. Right up to date, everything from dogs, cats, guinea pigs, various reptiles and parrots have all been used in homes, hospices, hospitals and prisons. Our need for animals is undisputed. While this kind of contact is very controlled, a kind of domesticated, stylised human-adapted version of wild, there are rawer, more true-to-roots forms of the same idea, simply promoting contact with nature. Like my robin.

Feeding the birds in the garden is a hobby for some fifty-five million people in the US, and in Britain it has been declared a national pastime for over half the adult population. Not surprisingly, it has frequently been adopted by many serving time with long prison sentences. Birds are symbolic of freedom, they are a distraction from the reality of life behind bars and a reminder that not everything in the world beyond the walls is denied them – those same therapeutic benefits of connection are in play again. The same goes for gardening and bee keeping; it's all about the same archaic stimulations that deliver a dose of hormones to create a deep sense of satisfaction that we can only get from nature.

I was rather pleased to discover that in 2016, the UK governmental advisory body on the protection of the environment, Natural England, published *A review of nature-based interventions for mental health care*. The fact that mental illness is on the rise, I suspect, may well be a symptom of the stresses and pressures of modern life. What better salve than the antidote of nature – my robin and my badgers sprang to mind simultaneously when I read the report.

Reading it reinforces what I've always known: simply being in nature reorders things, puts our own troubles into context, and gives us a rock to hang on to. When we take ourselves out for a walk in nature, our ancestral heritage whispers in our ears and fills our senses with what we are designed to be able to cope with; we feel valid again, part of the bigger process of life and, with that, we have a genuine reason to be alive. *Solvitur ambulando*, it is all solved with a walk.

Our National Nature Reserves in the UK are flagships – they were conceived to protect valuable habitats, either for their rare wildlife and biodiversity value or for their unique geology. Time is proving them to be a very sound investment; their value is appreciating and the dividends they return are way beyond those originally envisaged, and while their purpose to save species hasn't met with a perfect track record, the fact that we are still losing our wildlife, and the overall trend is worsening, bears witness to the need for them. They are still vital repositories of species from which, if we get our future priorities right, the bigger landscape can be seeded, something known as landscape scale conservation – a joining-up and reconnecting. In a way, something that could be defined as a kind of rewilding – somewhere between a lynx and a scruffy lawn – a sentiment very much like that expressed by those subscribers to North America's Yellowstone to Yukon initiative.

Despite the dysfunctionality of the bigger countryside, when you pass into a protected area, you immediately feel it;

the contrast with the outside is becoming greater, the difference is stark.

Many of us have come to know nature reserves and appreciate the 'hidden' value of wild that is so evident in these places. However, only when this value becomes the prescription does it make it to the mainstream.

Recognised by the health services, it makes its way onto political agendas and we have started to slowly take note. These nature reserves are undoubtedly important in conservation terms and it goes without saying that they are excellent places to experience nature in all of its wonderfully stimulating forms, and indeed to practise much of what this book has covered. But there is a change, an undercurrent of appreciation that has until now been missing. A number of reserves are hosting health initiatives such as Greencare and Ecotherapy, in realisation of the benefits that just being outside can bring to mental and physical wellbeing. In America, the identification of the negative health consequences of a lack of green space has led to the coining of the phrase 'Nature Deficit Disorder', something that is very real and for which the evidence is stacking up.

The current generation of young people is experiencing the highest levels of mental health and obesity issues to date. This creates a huge burden on the health system and, saddest of all, some say they are the first ever generation whose life expectancy falls short of their parents'. This is a real problem that isn't going away any time soon and it has been demonstrated beyond doubt that the quality of life of all is improved when we are given access to nature and green spaces.

*   *   *

It's not too much of a leap into the unknown to take this a step further. If nature is good for you, however you choose to experience it, why limit it to nature reserves? Why restrict your experiences to these parcels of paradise, which, fenced and gated, can sometime feel like commodities? 'Get your car keys, we're off to see some nature' isn't integration. It doesn't take that much of a paradigm shift to take it home with you. If nature is a medicine, don't just go and see the doctor, bring back home the drugs.

It's what my robin is to me. In fact, anything in my modest back garden that has come to be there without me buying it or putting it there directly constitutes my daily medicine and more. It is my sustenance, not directly, as it once was, although I'm not averse to a bit of foraging, but it is nutrition of sorts: just sitting out on a bright spring morning and watching the bee flies bowing to the flowers of lungwort, the unravelling of the ferns' fiddle heads, that ultraviolet shard of summer in a common blue damselfly . . . it's a pharmacy of countless coloured pills with all their natural uplifting powers. They're not mine and they're not yours; they're wild and they're free in every sense. Take as many as you can, you can't overdose on nature.

This is just a garden; imagine what it would be like if we started to join up a few more of the green dots on the landscape, linking the nature reserves, these repositories of the wild, with new ones, recreating a new future, schools, gardens, community, bit by bit. This realisation is to me the very beginning of rewilding at a level we can all take part in.

If the underlying concept of what is often referred to as ecological rewilding is about recreating fully functional ecosystems, then self-rewilding is about creating fully functional humans, free from ignorance and ecological prejudice, able to engage with the wild world and each other in a better, more compassionate way.

While we consider releasing long-missing beasts into our countryside in an effort to replace essential machinery in broken-down ecology, we are looking to rewild to recreate a better-functioning whole. We can also do with a similar human rehabilitation, for that is really what it's all about. The liberation of our own inner beast, recognising what it needs, our own essential requirements, space to unwind, rest, relax, reflect and forage, to heal and be inspired – these are factors that give us all a better quality of life and that is without considering the various other ecological services these places provide: flood-mitigation, clean water, shelter, breathable air, pollination – the list is literally endless, for we are not separate from this nature but part of it, and always have been.

I like to think of a future where this connection is taught: imagine an education system which makes outdoor time and environmental learning part of the national curriculum. This would be part of a human rewilding concept, a metaphorical Yellowstone wolf, about to create a cascade of positive change in our everyday life and culture. One that will have as much positive influence on its environment, and on our more ecologically functional future. One individual step at a time.

*   *   *

We all could do with feeling the difference nature can make, falling in love with it and finding a personal connection and discovering these positives. By becoming more aware of other life forms we develop a compassion for all. This liberating of old ideas and establishing of long-lost natural functions of our own bodies is rewilding too. Just as nature needs a coherent network of habitat connections through which to move, so does our inner wild; we need to reconnect to establish a different relationship with nature and the wild spaces we have left. In order to have any kind of a future this re-evaluation of the natural world and our place in it needs to happen very soon, before it is too late.

It is very subtle and it starts with personal awareness. It can be daunting to suddenly be expected to have an informed opinion on ecological rewilding, if you don't understand what's happening in your own window box, garden or park. How can you conceivably tolerate wolves at large in the landscape if you can't do your own thing by creating a space and habitat for a wolf spider?

We need to rewild from the bottom up, but also from the inside out. I would love to see a future in which we have understood the importance of, and are willing as a species to tolerate the concept of the reintroduction of, some of those keystone species, like lynx and beaver. It will happen (it's already happening), I'm sure, at some level. But in the meantime, the process of rewilding starts with us, our attitude to, and how we value, nature; we need to create or re-create those relationships, those intimacies, those feelings of respect that can only come from a better understanding of our environment. Because unless we rewild ourselves from within and find a profound connection between ourselves and nature and become re-enchanted with our world, no matter how hard we try, conservation efforts to date will be to no avail. It's time for us all to become more eco-savvy and, what's more, it's a lot easier than you might think.

Look up 'wild', and alongside it in the dictionary is a collection of not particularly inspiring synonyms; none of them seem that positive, do they? Who would want these words associated with them? If you're wild, you might be: undomesticated, untamed, ferocious, barbarian, unbroken, savage, tempestuous (OK, we all like a bit of that), turbulent, frenzied, self-willed, riotous, wayward, uncontrollable, reckless, rash, grotesque, bizarre, strange, unkempt.

The truth is that as a species, as soon as we could, we've tried to get away, control and manipulate the wild – and that includes these definitions. It's no wonder we have such an antagonistic relationship with much of wildlife on earth. Wild is not inherently bad; personally, I like to own many of these qualities. We need to embrace these words and make them our own in their original sense. If we dig into the words and their origins, we find other definitions and roots that temper the initial modern definitions. We start to find treasure.

This is a trove that is closer than you might think. The wild is a treasure that is priceless, yet we can all own it.

It is to be found everywhere you can turn a leaf, dig the dirt, see the sky and dip into the water. But we need to fast re-evaluate its stocks and shares, and realise its worth. This is the underpinning ideal behind rewilding in all its definitions – we need to treasure wild and all of its delicious unpredictability and experiences, and to do this we need to re-find what we've long lost, and make progress along a different path – and it all starts with you, right now.

Acknowledgements

This book would not have been possible without an incredible number of folk and various species who have helped, hinted, inspired and shared their thoughts and experiences with me over the years. I regret that their names are too numerous to thank all personally. However, special mention must go to the following.

My long suffering wife Ceri and my own little 'wild one' Elvie, who have had to tolerate a caffeine-crazed bear in the house for the months that this book was being hammered out on a keypad. Lucy Warburton who not only set this book in motion with a timely phone call, but has been the most gently persuasive and inspirational editor, one who has managed to pick me up and convince me to carry on in the face of monumental writer's block, and the spontaneous disruptions of my other parallel career. I must thank my copy-editor, Marian Reid, for her fantastic editing. I also had to take much nature therapy while this book was being 'built' and I must thank my new found friends in the New Forest, Martin Boxall and Ben Hobbs, for some restorative, long slow walks through the 'dimpsy' half-lit woods.

Thank you all.

Nick Baker is simply a naturalist. His life-long passion and natural curiosity for anything living has led to a varied career as a field biologist, broadcaster and communicator of natural history. Nick has over twenty years' experience hosting shows for the BBC, National Geographic and Discovery, including *Springwatch*, *Weird Creatures* and *Ultimate Explorer*. He has also authored ten books on the subject of wildlife and natural history. Nick is a strong believer in the fact you can't protect what you don't value or connect with, and so, to this day he continues in whatever way he can to help others rediscover a natural empathy with the wild world.